# RLT Approaches to QSAPs

—

## Applied to Timetable Synchronization in Public Transport

Beim Fachbereich Mathematik
der Universität Kaiserslautern
zur Erlangung des akademischen Grades
Doktor der Naturwissenschaften
(Doctor rerum naturalium, Dr. rer. nat.)
eingereichte

### Dissertation

von

### Ingmar Schüle

February 2010

Bibliografische Information der Deutschen Nationalbibliothek

Die Deutsche Nationalbibliothek verzeichnet diese Publikation in der
Deutschen Nationalbibliografie; detaillierte bibliografische Daten sind
im Internet über http://dnb.d-nb.de abrufbar.

ISBN 978-3-8325-2637-5

Logos Verlag Berlin GmbH
Comeniushof, Gubener Str. 47,
10243 Berlin
Tel.: +49 (0)30 42 85 10 90
Fax: +49 (0)30 42 85 10 92
INTERNET: http://www.logos-verlag.de

*Coming back to where you started
is not the same as never leaving.*

Terry Pratchett - A Hat Full of Sky

# Acknowledgements

This work was done with the financial support of the Department Optimization of the Fraunhofer Institute for Industrial Mathematics (ITWM).

I would like to express my sincere gratitude to my supervisor PD Dr. habil. Karl-Heinz Küfer for his support, guidance and encouragement throughout the last years. Special thanks go to him and Dr. Michael Schröder for guiding me and for introducing me to this challenging research topic.

I would also like to thank the VRN and Frieder Zappe for supporting me with real-life data and for creative discussions about the topic of timetable synchronization. Special thanks go to my colleagues of the SynPlan team, Alex, Anca, Elena, Faisal, Michael, Neele and Yavor, for the great support and the excellent working atmosphere in the last three years.

I would especially like to thank Hendrik for countless creative debates and Gesche, you are a great listener.

For proof-reading and many helpful comments, I thank Dieke, Hendrik, Martin, Michael K., Neele, Rebekka, Sebastian and Tabea. In addition, I want to thank Clarisse, Tobias and Uwe for their contribution to this work.

Finally, I want to thank my family for their love and their great support during my studies.

# Contents

# Chapter 1

# Introduction and Motivation

The research conducted in this thesis has mainly been inspired by the need to synchronize timetables in public transport systems. In this section, we introduce the concept of and a motivation for the Timetable Synchronization Problem (TTSP), present the outline of this thesis and describe the mathematical problems and challenges that have occurred during the research.

### Problem Description

Surveys of the behavior of travelers show that the inconvenience of changing from one means of transport to another one is a decisive reason not to use public transport systems (cf. [3]). Thus, in times when environmental pollution is an important issue and when the scarcity of raw materials is a critical topic for the near future, it is important to improve public transport systems so that passengers can transfer more conveniently.

In general, constructing a good public transport network is a process that contains several optimization problems. The bus stops must be placed so that passengers do not have to walk long distances, the routes of the bus lines must be planned to connect these stops in appropriate ways and the starting times of the single vehicles must be scheduled. Here, periodic structures as well as aperiodic starting times may be considered, depending on passenger demand. Once all of these optimization problems have been solved, the need for synchronizing the timetables of different companies arises.

In most urban areas, several companies provide public transportation and each company operates its own fleet of vehicles. Long-distance trains connect distant cities, regional trains and regional buses connect the suburbs

Figure: Passengers at the main station of Kaiserslautern.

with the city center and city buses and subways transport passengers within the city. If several companies are involved, a public transport association is often responsible to standardize and unite the different ticketing and pricing systems. While the timetables of the companies are internally synchronized, there is a lack of synchronization at the intersection points of the different subnetworks. Here, the planners of the public transport associations have to deal with the problem of finding good synchronizations.

Public transport systems are grown structures that are highly interconnected. This situation makes it nearly impossible for the traffic planner to have an adequate overview of the network-wide consequences that local changes will cause. Therefore, computer based decision support systems are needed to create a better synchronization in the traffic networks. In addition, trying to improve all transfer possibilities in a network is a hopeless task due to the huge number of transfers and the high interconnectedness of the network.

Fortunately, the majority of passengers change lines only at certain important points in the network. Here, the quantity of passengers is comparatively high. Examples of such important network nodes are, in most cities, the main station and the city center. Focusing the analysis on these places in the network results in a much more efficient process of timetable synchronization than regarding the whole network with its unmanageable transfer possibilities.

To influence the given timetables, we allow changes of a few minutes to the starting times of the public transport lines. Our goal is not to create totally new timetables but to apply small changes to the current ones to generate better transfer possibilities for the passengers. An advantage of shifting whole lines is that most vehicle schedules are maintained. Here, shifting all starting times of the tours of a single line does not impair the whole vehicle schedule of a company.

Former research on the topic of timetable synchronization was done in the late eighties, but the approaches never found their way into practice. One reason for this outcome is that the given methods used the simple strategy of minimizing the waiting time for the passengers. But such a consideration has the side-effect that the generated small waiting times also cause the risk of missing a transfer if the arriving vehicle is delayed. Therefore, the goal of our approach is to generate transfers that can be called convenient: The waiting time for the passengers should neither be too short, resulting in a risky transfer, nor too long, so that passengers need a lot of patience while they wait for their transfer. All in all, we split the waiting time of a transfer into six intervals that are presented in Figure 1.1.

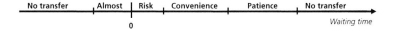

Figure 1.1: Classification of the waiting time for the passengers.

In addition to the intuitive risk, convenience and patience intervals, we also focus on short negative waiting times. These almost transfers are interesting for the traffic planners, since they correspond to very unsatisfying situations for the passengers. Furthermore, they can be easily transformed into real transfers by small changes.

In general, the TTSP has a multiobjective character. Every single transfer presents an optimization goal on its own. Thus, the trade-offs between improvements and changes for the worse of the transfers must be considered in the objective functions.

The knowledge of the traffic planner has a local character that is related to single transfers. In contrast to this, our optimization approach works on a global level as it tries to improve the network-wide situation. This thesis introduces an interactive process, where the traffic planner sets constraints on a local level to influence the solutions of the global algorithms. The planner inspects a series of optimized timetables and sets new constraints to influence the optimization so that the generated timetables of the following iteration fit his needs better. This interaction proceeds until a set of timetables is created that fulfill all requirements.

### Outline of the Thesis

The research of this thesis is motivated by the TTSP. For this problem, we need new mathematical methods that are able to solve it efficiently.

Voss [62] modeled the TTSP in 1992 as a Quadratic Semi-Assignment Problem (QSAP). A QSAP is the task to assign $m$ objects to $n$ locations (not necessarily one-to-one) so that a quadratic objective function is minimized. The problem is known to be NP-hard and most of its real life applications cannot be solved by exact solution strategies in a reasonable amount of time.

The QSAP that arises from the TTSP has a more general structure than the one that is most often regarded in literature. In our case, the feasible locations may differ for the individual objects. For a better understanding of this generalized QSAP, we extend a polyhedral formulation that is based on the work of Saito et al. [49] from 2004.

Mathematical solvers can compute optimal solutions of the Mixed Integer Program (MIP) formulation that belongs to the QSAP, but the computation times are far too long for problem sizes of real life applications. In addition, this technique does not take the multi dimensional character of the TTSP sufficiently into account. It is important to realize that there does not exist a timetable for a public transport network that can be called "optimal". Trade-offs between improvements and worsenings of the situation at different network nodes cannot be rated in a single objective function. Thus, multi-criteria approaches must be used to generate solutions that have good characteristics for the different goals of the optimization.

We apply modern metaheuristics to the problem to generate this set of "good" solutions quickly. This thesis presents a new multi-colony Ant Colony Optimization (ACO) approach in detail, where several ant colonies try to solve the problem by sharing information via pheromone trails. Single ants construct different timetables and promising combinations of changes are marked with higher pheromone values. This type of algorithm is able to find good solutions for the related Quadratic Assignment Problem (QAP). Thus, applying it to the TTSP seems to be promising. In addition, we also apply a Genetic Algorithm and a Simulated Annealing approach to the QSAP.

A general problem of metaheuristics is that there is no certainty about the quality of the solutions. Hence, an approach to generate lower bounds for the solutions is needed. Trivial bound techniques and a simple linearization of the problem are useful to generate weak lower bounds quickly, but the gaps of these bounds are too large for an efficient analysis of the solution quality of the metaheuristics.

Recent improvements in the research area of lower bounds for QAPs were achieved with the help of the Reformulation Linearization Technique (RLT) (cf. [1], [24] and [56]). The RLT is an approach that generates stepwise tighter polyhedral formulations of the QAP by introducing cutting planes in higher dimensional spaces. We can use LP solvers to solve the resulting linear formulations and to generate solutions that provide lower bounds for the QAP.

We analyze in this thesis the RLT for the related QSAP. Here, the stepwise structure results in formulations that are either fast to compute with high gaps or that return good bounds but only with high computational efforts. Thus, we introduce new ideas to break up the stepwise formulation of the RLT. These new considerations further improve the applicability of this technique. It was up to now unknown for the stepwise improving RLT formulations from which step on they are exact, meaning that the lower bounds are equal to the optimal solution value of the QSAP. We prove a tight polyhedral characterization of the RLT formulation, which provides the minimal RLT level that is needed for an exact formulation of a certain QSAP size.

A topic that is related to the RLT is the research area of Pseudo-Boolean Optimization. Here, a reduction technique allows changes to the objective function without altering the structure of the QSAP. With these changes, we can extract a constant value from the objective function that can be used as a lower bound of the QSAP. This technique has the advantage that it does not depend on external LP solvers and that it generates good bounds very fast in comparison to the MIP or the LP formulation.

We test the generated approaches on various QSAP instances. The focus of the test runs lies on the quality of the lower bounds and the total time that is needed for the calculation. In addition, we have a look at the progress that the algorithms make during their runtime for an efficiency analysis. The lower bounds are used to evaluate the practicability of the ACO metaheuristic for the TTSP.

The new theoretical results of this thesis are achieved due to questions that arise from real life applications. The new approaches are implemented in a software concept that is used to improve the public transport network of the city of Kaiserslautern. Further applications are planned in the near future.

The outline of this thesis is presented in Figure 1.2.

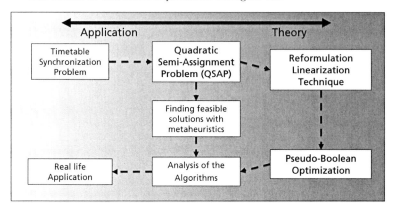

Figure 1.2: Outline of the thesis. The topics in bold indicate the mathematical focus of this work.

In Chapter 2, the TTSP and our approach to model it are presented in detail. Here, all important aspects of the problem are discussed and the goals and objectives of the traffic planners are illustrated.

Chapter 3 introduces the QSAP and some of its applications. The problem is modeled as an MIP formulation and we also present a relaxed LP formulation. Furthermore, we discuss the connection to the related QAP. For this, the best known lower bound approaches are a Semidefinite Programming approach and the RLT. We discuss the applicability of both approaches for our generalized QSAP. For a better understanding of the problem, we extend a polyhedral analysis so that it describes our generalized QSAP that is used for the TTSP. Finally, we present a simple lower bound approach for the QSAP.

The subsequent Chapter 4 presents metaheuristic solution strategies for the QSAP. Three different methods, namely a Simulated Annealing approach, a Genetic Algorithm and an Ant Colony Optimization method are applied to the multi-criteria QSAP. From these three approaches, the multi-colony ACO algorithm is presented in detail. The algorithm is hybridized with a Local Search approach to increase its efficiency.

Since metaheuristics only generate approximated solutions, we introduce an RLT approach in Chapter 5 that can be used to evaluate the approximation quality. This chapter contains the most relevant mathematical innovations of this thesis. We extend the existing RLT approach by breaking up the stepwise structure of the RLT. For this, we introduce an RLT-1.5 algorithm that improves the RLT-1 formulation by successively applying RLT-2 constraints. Thus, the algorithm presents a trade-off between these two formulations. In addition, we extend the theory of the RLT for QSAPs by proving a novel minimal tight RLT formulation result.

Furthermore, we present another efficient lower bound approach for the QSAP that is based on a reduction technique. This approach, which is based on the theory of pseudo-Boolean Optimization, computes lower bounds of the QSAP by extracting a constant term from the objective function. A hybrid approach of this reduction technique and the RLT-1.5 algorithm provides remarkable bounds for the QSAP test instances.

Chapter 6 analyzes and compares the efficiency of the different approaches from the previous chapters. The algorithms are tested on real life data from the city of Kaiserslautern as well as on randomly generated test instances. Here, we also compare the results of the ACO metaheuristic with the solutions of the MIP formulation and the lower bound approaches.

Finally we discuss the application of the model and the approaches on a real life example for the city of Kaiserslautern in Chapter 7. Here, the strengths and weaknesses of the model and the objective functions are discussed. Furthermore, we give a brief illustration of the software tool that provides decision support for the traffic planner for synchronizing timetables.

## Mathematical Problems and Challenges

This work is based upon a real life problem. Therefore, a realistic model of the actual processes in public transport networks is necessary. This includes a model of the behavior of the passengers. Furthermore, it is hard to find a good trade-off between the multiobjective structure of the problem (every transfer should be optimized) and a manageable amount of objectives for the optimization process.

In addition to the difficulties that arise from the modeling part of the thesis, there are also a lot of mathematical challenges. The QSAP is known to be NP-hard and it is unrealistic to try to solve larger QSAPs exactly in an acceptable time, even if only one objective function is considered. The QSAP is not widely studied and there are no results known yet for the multiobjective constrained QSAP. Hence, we have to analyze different solution strategies for their applicability to our problem.

The results of the metaheuristics that we apply to the QSAP have the disadvantage that the solution quality is unknown. To evaluate the approximated Pareto frontier, lower bound approaches can be used. Such bounds are presented in Figure 1.3, where we evaluate the solutions of the ACO metaheuristic with the help of lower bounds.

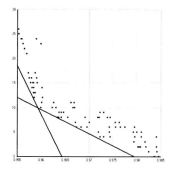

Figure 1.3: Visualization of solutions of the TTSP that are generated by the ACO metaheuristic. Each point represents an improved timetable. The $x$-axis shows a penalty value that indicates the quality of a timetable and the $y$-axis presents the degree of changes that is made to the current timetable. The goal is to minimize both values. The bold lines represent lower bounds and no solutions can exist bottom left of these lines. The bounds can be used to evaluate the quality of the solutions.

For these lower bounds, ideas from the QAP can be used and adapted so that they also work for the QSAP. The RLT is currently one of the best known methods, but our analysis shows that the RLT-1 bounds are too weak and that we need too much time to compute the RLT-2 bounds. Our new approach combines the simplicity of the RLT-1 formulation and the strength of the RLT-2 formulation by successively improving the formulation.

A different approach for generating lower bounds for the QSAP is to reformulate the objective function. The height $H_{\mathcal{A}}[f]$ is the maximum constant value $c$ that can be extracted from the objective function $f$ so that there exists a homogeneous quadratic posiform $\phi(x, \bar{x})$ $(\bar{x} = 1 - x)$, for which $f = c + \phi(x, \bar{x})$ holds for all feasible assignments $x \in \mathcal{A}$ of the QSAP. We extend this formulation to a stepwise tightening formulation $H_{t,\mathcal{A}}[f(x)]$, where $t$ defines the degree of the posiform $\phi(x, \bar{x})$. This new definition of the height is strongly related to the RLT for QSAPs and we present first results on this topic. By approximating the height with a Dual Ascent algorithm, a lower bound for the QSAP can be generated. Here, we can also apply the break up strategy from the RLT-1.5 algorithm to generate an approximation between different height levels.

Finally, we present a hybrid approach that combines the efficient non-deterministic Dual Ascent algorithm with the stepwise tightening RLT-1.5 formulation. This hybrid technique provides the best generated bounds for the bigger real life test instances.

# Chapter 2

# The Timetable Synchronization Problem

The research in the area of public transportation contains a lot of different optimization problems. Track laying must be optimized so that routes are short, stations should be placed well-considered to cover a huge area, the detailed routes for buses must be planned and timetables are needed so that all lines can be served. This thesis and its approaches start after all these optimization problems are already solved in a public transport network.

The Timetable Synchronization Problem (TTSP) deals with the task to coordinate the public transport system of an area not by creating a new timetable but by applying small changes to the current one. This demand for synchronization mainly occurs, when several companies interact within the same network. Although the company internal timetables are good synchronized in the majority of cases, there is often the need for a better coordination at the intersection points of the different companies.

Some scientific work was done on this kind of problem in the late eighties, but the results never found their way into the real life process of timetable synchronization. In our opinion, this was due to two main reasons. Firstly, the chosen approach of simply minimizing waiting times does not generate transfers that can be called convenient from passenger perspective. This is the case, because very small waiting times also have the negative aspect of a high risk of missing a transfer. And secondly, the more important reason is that the multi-criteria aspects of the problem were not taken into account according to their importance.

This chapter describes our approach of modeling the TTSP, the goals of the traffic planners and the general requirements that must be considered while optimizing timetables.

## 2.1   Literature Survey

The topic of optimization in public transport contains many different optimization problems. An example is the work of Schoebel [50], which treats the stop location problem, the question of delay management and a tariff planning approach.

This thesis deals with the optimization problem of timetable and schedule synchronization in public transport networks. In the late eighties and the beginning of the nineties, a lot of work was done concerning this topic. The first publications were made by Klemt and Stemme [33] in 1988 and Domschke [13] in 1989. Here, the problem was formulated for the first time and some heuristic approaches were tested for several given test instances. The goal was to minimize the average waiting time for all passengers.

A few years later in 1992, more detailed analyses were made by Bookbinder et al. [7], Désilets et al. [11] and Voss [62]. These papers enlarge and refine the given model and the used approaches from the earlier works. But none of these works found, to the best of our knowledge, its way into practice.

In the late nineties (1997), Adamski and Byniarska [2] applied some new heuristics, namely Tabu Search and a genetic method to the problem and in 2004, Fleurent et al. [16] changed the concept of minimizing the waiting time into an approach of creating good waiting times.

The first work that found its way into practice was for a special case of the TTSP that was optimized by Liebchen [35] in 2005. Here, the timetable of the underground railway of the city of Berlin was improved. One main aspect of this work is that the underground drives in periodic times. The work is a part of a PhD-thesis about the optimization of periodic timetables [36]. It is based on the basic graph model of the Periodic Event Scheduling Problem (PESP).

Our work on the topic of timetable synchronization started with a general feasibility study in 2006 by Schroeder and Solchenbach [52]. Here, ideas from literature were enhanced and some new approaches were constructed. The results and ideas of this work were afterwards refined in the diploma thesis of Sonja Klein [32].

During our research, we published some non-mathematical publications on the topic of timetable synchronization, namely a publication in the journal Straßenverkehrstechnik in 2007 by Schroeder and Schuele [51] and an article in the conference proceedings of the Urban Transport conference 2009 in Bologna [55]. The problem of introducing the vehicle scheduling into our model was analyzed in the diploma thesis of Neele Hansen [26].

## 2.2 Description of the Problem

Public transport networks of bigger cities are mostly run by more than a single company. Trains transport the passengers from other urban areas or cities to the central station, passengers take city buses or if available trams within the city center and for passengers from the suburbs, regional buses provide a good connection to the city.

In most urban areas, these different parts of the network are run by different companies. Such a company has its own optimization strategy to find good timetables for its own network. This includes the planning of human resources and the vehicle scheduling, but seldom are the timetables of the other companies considered sufficiently in this process. Here, the TTSP gets involved. Especially at the intersection points of the different companies, a lack of synchronization can be found. The challenge is now to find changes that can be applied to the system that improve the synchronization at the intersection points and, at the same time, do not destroy the company-internal plannings, such as the vehicle scheduling.

But what is a good synchronization? And how can such complex things like timetables be compared? These questions are two of the main issues in this chapter.

## 2.3 Main Ideas of our Approach

Our concept of synchronizing timetables improves the convenience of the passengers and not necessarily the earnings of the companies. These days, in times of ecological and financial problems, more and more people start to think about using public transportation. But, compared to traveling by car, it is uncomfortable to take the bus and long waiting times or the risk of missing a transfer prevents the passengers from changing the traffic mode they are used to.

Here, our approach starts by changing the standard concept of minimizing waiting times into a concept of maximizing the convenience of the passengers. Short waiting times are risky, long waiting times are annoying, so a good compromise lies somewhere in between. In addition, we introduce the concept of "almost transfers". Such an almost transfer is not a real transfer, since it refers to a situation, in which the departing vehicle has already left a short time before the passenger arrives at a station. This situation is not caused by a delay, but by a lack of synchronization. Hence, by finding these almost transfers, we have a starting point for detecting good synchronization possibilities, since small changes can easily change these transfers into existing transfers.

Synchronizing timetables is a very complex problem and it is not reasonable to model it in a one-dimensional way. This is the case, since it is hardly possible to express the diversity of the problem in a single objective function. Instead, we create a multi-criteria model that combines both, the global view on the whole network and the local view on the single transfers. This diversity is integrated in an interactive optimization process, where mathematical algorithms optimize the global situation with respect to constraints that the traffic planner defines on a local level.

Since the whole network contains far too much information (and most of it is irrelevant for the TTSP), we concentrate our analysis on important transfer locations in the network. These locations, called network nodes, do not necessarily consist of a single station but they can also be a set of stations where passengers are willing to walk a small distance to reach a departing line. This focusing on the important places in the network gives the results a higher significance compared to an analysis, in which every possible transfer is regarded. It also reduces the problem size to a level that can be handled by modern computer systems.

The variable elements of our model are the starting times of the tours. But to preserve the vehicle schedules and the periodic structure of the timetables, we only allow changes to the starting times of whole public transport lines. This approach is chosen to maintain the given vehicle schedule, since, in the majority of cases, vehicle schedules are based on single lines. This means that most vehicles drive only tours of the same line each day.

## 2.4 Modeling

In this section, the important definitions that are related to the TTSP are introduced. For a clear structure, we separate them into definitions that are related to

- the infrastructure of the public transport network,

- the transfers between vehicles,

- the evaluation of the transfer quality.

### 2.4.1 Modeling of Infrastructure

Infrastructural definitions refer to the classical terms that are public transport related. The geographical aspects are the bus stops and the train stations, where passengers alight from vehicles or get on vehicles.

**Definition 2.1 (Station, Network Node)** *A station is a location where passengers board vehicles or get off vehicles. In this work, we refer to a station by $\omega$. The set of all stations is $\Omega$. A network node $\delta$ is a set of geographically nearby stations, where passengers are willing to walk small distances to change between vehicles. $\Delta$ denotes the set of all network nodes.*

A simple form of a network node is, for example, a bus station that is in the neighborhood of a train station. Here, passengers can walk a short distance for a transfer from a bus to a train or vice versa. But combining some stations to a network node can also be useful within the stations of a single company.

A central aspect of the timetables are the tours of the single vehicles. A tour describes the route of a bus or a train and includes all stations that a vehicle stops at and all corresponding times (arriving time and departing time) for these stations. Such a timetable for three tours can be seen in Table 2.1. In most cases, only the departing times are presented in public timetables. This is done to avoid confusions for the passengers. Internally, the arriving times and the departing times are separated, since it is very important to have the exact arriving times while regarding transfers.

|            | tour 1 | tour 2 | tour 3 |
|------------|--------|--------|--------|
| station 1  | 07.15  | 07.45  | -      |
| station 2  | 07.18  | 07.48  | -      |
| station 3  | 07.21  | -      | 08.21  |
| station 4  | 07.24  | -      | 08.24  |
| station 5  | 07.27  | 07.55  | 08.27  |
| station 6  | 07.30  | 07.58  | 08.30  |
| station 7  | 07.34  | 08.02  | 08.34  |
| station 8  | 07.36  | 08.04  | 08.36  |
| station 9  | 07.40  | 08.08  | 08.40  |
| station 10 | 07.42  | 08.10  | 08.42  |
| station 11 | 07.47  | 08.15  | 08.47  |
| station 12 | 07.50  | 08.18  | 08.50  |
| station 13 | 07.52  | -      | 08.52  |
| station 14 | 07.55  | -      | 08.55  |
| station 15 | 08.00  | -      | 09.00  |

Table 2.1: Example of a simple timetable. Only departing times are presented.

**Remark 2.2** *Note that all times that are regarded in this thesis are discrete and integer. Timetables in public transport that are based on seconds are unrealistic and not desirable. Thus, all times in thesis are regarded in minutes.*

The following definition introduces the mathematical notation for the concept of tours.

**Definition 2.3 (Tour)** *A tour $\zeta$ is an ordered set of triples*

$$(\omega, t_{arr}(\omega), t_{dep}(\omega)),$$

*where $\omega$ is a station, $t_{arr}(\omega)$ is the arriving time and $t_{dep}(\omega)$ is the departing time at this station. We define the set of all tours by $T$.*

All tours of a company are bundled in lines that share equal or at least similar routes. In most cases, the tours of a line stop at the same stations in the middle of the route and differ only at the first few stations or the last few stations (cf. Table 2.1). The exact definition of a line is as follows.

**Definition 2.4 (Line)** *A line $l$ is defined by a set of tours $l = \{\zeta_1, ..., \zeta_n\}$. Thus, it is a subset of the set of all tours ($l \subseteq T$). $L$ denotes the set of all lines and it is a partition of $T$.*

To bundle those lines that are strongly related inside the network, we introduce the concept of line clusters. This relation concerns vehicles, drivers, infrastructure and the whole timetable planning process. A simple example of such a clustering is to bundle all lines that belong to the same company.

**Definition 2.5 (Line Cluster)** *A line cluster $\lambda = \{l_1, ..., l_s\}$ is a bundle of lines that share similar characteristics. For the set $L$ of all lines, a partition $\Lambda = \{\lambda_1, ..., \lambda_t\}$ of $L$ denotes the set of all line clusters.*

For the relation between the tours and the lines of a timetable, we have

$$\bigcup_{l \in L} \bigcup_{\zeta \in l} \zeta = T \quad \text{and} \quad \bigcup_{\lambda \in \Lambda} \bigcup_{l \in \lambda} l = L.$$

In most cases, a line can be regarded as a route between two places. For two stations $\omega_1$ and $\omega_2$ that are connected by a line, we have the tours that drive from $\omega_1$ to $\omega_2$ and the tours with the opposite direction. To differ between these tours, we introduce the definition of a direction.

**Definition 2.6 (Direction)** *Each tour of a line has a unique direction $\rho$. In this work, we will refer to these directions as 'forward' and 'backward', so $\rho \in \{forward, backward\}$.*

In some rare cases, a line can be a circular trip that is only used in one direction. In this case, all tours of the line get the direction 'forward' and the set of tours of this line with the direction 'backward' is empty.

## 2.4.2 Modeling of Transfers

It is the main aspect of the TTSP to make transfers between vehicles as convenient as possible for the passenger. Thus, an appropriate definition of transfers is needed. They are the central element in our model, where we measure the convenience of passengers.

**Definition 2.7 (Transfer)** *A transfer $\tau$ is a quadruple $\{\delta, \zeta_{arr}, l_{dep}, \rho_{l_{dep}}\}$ with a network node $\delta \in \Delta$, an arriving tour $\zeta_{arr} \in T$, a departing line*

$l_{dep} \in L$ with $\zeta_{arr} \notin l_{dep}$ and a direction for the departing line $\rho_{l_{dep}} \in$ {*forward, backward*}. *We denote the set of all transfers by* $\mathcal{T}$.

To bundle the transfers that share similar characteristics, we introduce the concept of connections. Here, we group those transfers that refer to the same lines and directions. Thus, a connection is defined by a network node, an arriving line with a direction and a departing line with a direction. In addition, we introduce the concept of connection types. A connection type contains all connections between two line clusters at a network node.

**Definition 2.8 (Connection, Connection Type)** *A connection $c$ is a quintuple $\{\delta, l_{arr}, \rho_{l_{arr}}, l_{dep}, \rho_{l_{dep}}\}$ with a network node $\delta \in \Delta$, two public transport lines $l_{arr}, l_{dep} \in L$ ($l_{arr} \neq l_{dep}$) and two directions $\rho_{l_{arr}}, \rho_{l_{dep}}$. The set of all connections is denoted by $C$.*

*A connection type is a triple $\{\delta, \lambda_{arr}, \lambda_{dep}\}$ and refers to the meeting of two line clusters $\lambda_{arr}$ and $\lambda_{dep}$ at a network node $\delta$.*

Since every tour $\zeta$ belongs to a single line $l$ and has a unique direction $\rho$, each transfer $\tau$ belongs to a unique connection $c$.

**Remark 2.9** *While constructing the set of all transfers $\mathcal{T}$, an important decision has to be made. Generally, there are two main events that can cause a transfer. One is the arrival of a tour in a network node and the other one is the departure of a tour.*

*Both approaches have their advantages and disadvantages. In this work, we decided to focus on a forward analysis, where passengers plan their tours from the starting point to the destination point. This corresponds to the view that arriving tours cause transfers. The opposite way would be to plan tours from the destination backwards to the starting point, but such a backward analysis concept would be confusing for most passengers.*

*Note that not every arriving tour has an associated departing tour. As a counterexample, we can regard a high-frequency arriving line and a low-frequency departing line. Here, it is not possible that all arriving tours generate good transfers.*

*An example of a matching of arriving tours to departing tours is presented in Figure 2.1.*

With this definition of transfers, there are still some open questions about the behavior of passengers. There are choices like:

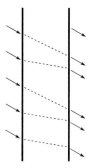

Figure 2.1: Transfers generated by arriving tours (arrows on the left side) are allocated to departing tours (arrows on the right side). The time is presented in vertical direction.

- at which station in a network node does the passenger alight?

- at which station does he board into the departing line?

- which departing tour does he choose?

For getting a suitable model, it is very important to have an appropriate description of how passengers actually transfer. To describe these choices properly, we introduce the following definitions.

**Definition 2.10 (Transfer Time)** *The time that a passenger spends in a network node while transferring between two lines is called transfer time $t_{trans}$. It is defined as the time difference between the first arriving time of the arriving tour at a station of the network node and the last departing time of the departing tour at a station of the network node. The value $t_{trans}$ is fixed for each pair of tours $\{\zeta_{arr}, \zeta_{dep}\}$, unless the arriving time or departing time of one of the lines is changed.*

For a better understanding of the term transfer time, we split this time up into three smaller time periods.

**Definition 2.11 (Driving Time, Walking Time, Waiting Time)** *A transfer time $t_{trans}$ can be split up into the driving time $t_{drive}$ that the passenger spends in vehicles, the walking time $t_{walk}$ that is needed to walk between* *two stations and the waiting time $t_{wait}$ that the passenger spends at the station, where he boards the departing tour. For these four times, we have the equation*

$$t_{trans} = t_{drive} + t_{walk} + t_{wait}.$$

*The walking time can be described by a predefined function with four parameters*

$$\mathcal{W} : \Omega^2 \times \Lambda^2 \to \mathbb{Z}$$

*that maps an arriving station, an arriving line cluster, a departing station and a departing line cluster to a walking time.*

Note that these four parameters are necessary, since, e.g., the location where the buses stop and where the trains stop at the same station can have a significant walking distance.

With the given definitions of walking time and waiting time, we can define how the passenger actually transfers between an arriving tour and a departing tour.

**Definition 2.12** $(\xi(t_{arr}, t_{dep}))$ *For an arriving tour $\zeta_{arr}$ and a departing tour $\zeta_{dep}$ at a network node $\delta$, the function*

$$\xi_\delta : T \times T \to \mathbb{R}, \quad \xi_\delta : \{\zeta_{arr}, \zeta_{dep}\} \mapsto t_{wait}$$

*maps these two tours to a corresponding waiting time. This waiting time is determined by the best arriving station $\omega_{arr}$ where the passenger steps off the arriving tour and the best departing station $\omega_{dep}$ where the passenger boards the departing tour. The stations $\omega_{arr}$ and $\omega_{dep}$ are chosen so that the waiting time $t_{wait}$ is maximized. This corresponds to a minimization of the sum $t_{drive} + t_{walk}$, since the total sum $t_{wait} + t_{drive} + t_{walk} = t_{trans}$ is constant for a fixed departing tour $\zeta_{dep}$.*

**Remark 2.13** *The maximization of the waiting time corresponds to a minimization of the risk of missing the departing tour in case of a delay of the arrival. Thus, such a choice is desirable for the passenger. This consideration is equivalent to minimizing the time that is needed to transfer inside the network node.*

*Note that there is not only one possible departing tour for each arriving tour. Therefore, we have to find for each transfer the optimal tour of the departing line (with the correct direction). Optimal means here that the waiting time $t_{wait}$ is as convenient as possible for the passenger.*

To evaluate transfers, we have to find a way to describe the convenience of passengers mathematically. This is done in the following subsection.

## 2.4.3  Evaluation of Transfers

For being able to improve the transfer quality, it is necessary to have an evaluation function that determines the quality (or the convenience) of a transfer. We estimate this convenience in our approach by regarding the waiting time of the passengers. One of the main improvements of the given model is that the waiting time is not simply minimized. Instead, we try to achieve a waiting time that can be called convenient. For this, we make a partitioning of the waiting time in the five transfer types that are presented in Figure 2.2. These waiting time classes are described in the following definition.

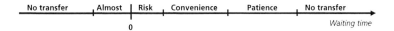

Figure 2.2: Waiting time classification in five transfer types and six intervals. The goal is to make all transfers as convenient as possible.

**Definition 2.14 (Waiting Time Classification, Transfer Type)** *A waiting time classification defines a partition of the waiting time in five different transfer types. These transfer types are:*

- *Convenience transfer: A waiting time that is not too long and not too short is called "convenient". It secures that small delays do not cause a missed transfer,*

- *Patience transfer: If a waiting time is too long to be called convenient, but still can be seen as a useful transfer, we call the transfer "patient",*

- *Risk transfer: A very short waiting time that causes the "risk" to miss the given transfer if the arriving tour is delayed,*

- *Almost transfer: A small negative waiting time is called an "almost" transfer,*

- *No transfer: We speak of a "no transfer", if no departing tour fits to an arriving tour in a relevant time interval.*

*A waiting time classifications is defined by the beginning and the ending of all time intervals.*

Note that the definition of "almost transfers" is useful, since they describe a very unsatisfying situation for the passenger. Furthermore, "almost transfers" can be easily transformed into real transfers by small modifications that are made to the timetable. Thus, they are good starting points for optimization approaches. A "no transfer" does not necessarily refer to a bad synchronization, since, e.g., a connection between a high-frequency line and a low-frequency line causes automatically a lot of "no transfers". A waiting time classification must be predefined for each combination of network node, arriving line cluster and departing line cluster. The classification for "no transfers" consists of two half-open intervals.

With a given waiting time classification, it is easy to formulate the goal that all transfers should become "convenience transfers". But since such a goal is impossible to achieve, the need for comparing transfer qualities occurs for making good decisions.

**Definition 2.15 (Penalty Function)** *A penalty function*

$$P : \mathbb{R} \to \mathbb{R}, \quad P : t_{wait} \longmapsto x$$

*is a function that maps waiting times to penalty values so that all waiting times of the same transfer type get the same penalty value. Thus, $P$ is defined as a step function. If required, different penalty functions can be defined for all network node, arriving line cluster, departing line cluster triples $(\delta, \lambda_{arr}, \lambda_{dep})$.*

An example of such a penalty function is presented in Figure 2.3. In general, a convenience transfer is the transfer with the lowest penalty and an almost transfer has the highest penalty. One can argue about the relation of risk and patience transfers, but both transfer types are certainly better than a "no transfer".

Using the penalty function from Definition 2.15 is a first approach to evaluate transfers. But it also has some disadvantages, e.g., a long risk waiting time should be preferred over a long patience waiting time. Additionally, the

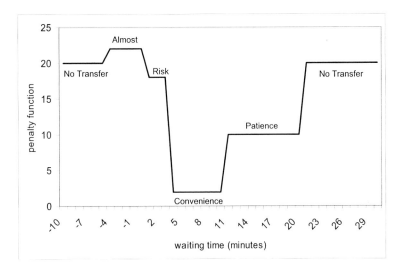

Figure 2.3: Example of a penalty function for a waiting time classification.

large differences of the penalties at the borders of the intervals are not desirable. Thus, we improve the penalty function with the help of a smoothing method.

**Definition 2.16 (Smoothed Penalty Function)** *For a given penalty function $P$, $\widehat{P}$ denotes a smoothed version of this function. Such a smoothed function avoids big differences between the penalties of adjacent waiting times.*

**Remark 2.17** *There are different approaches for generating a smoothed penalty function. We can use, e.g., cubic splines or other different methods. In our work, we decided to keep the penalties as integer values. So we implemented a special method that cubically smoothes the penalty functions. The approach works similar to cubic splines. But we take in addition the risk penalty as a preset value for the time point that is exactly in the middle between the end of the almost interval and the beginning of the convenience interval. The same is done for the patience penalty, which we take as a preset penalty value for the time point that is in the middle between the end of the convenience interval and the beginning of the no transfer interval. The* *generated penalty values are rounded to the next integer value to keep the penalties integral. We present an example of a cubically smoothed penalty function in Figure 2.4.*

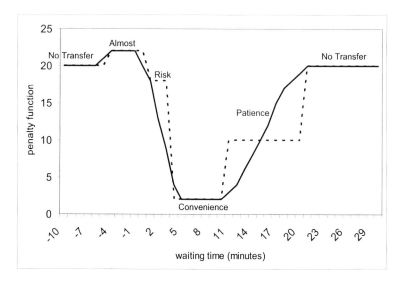

Figure 2.4: Example of a smoothed penalty function for a waiting time classification. The dashed lines represent the unsmoothed penalty function.

Up to now, we have not incorporated a very important factor in the given model: the passenger numbers. For public transport networks, it is very important to differ between tours that have high passenger numbers and those with only a few passengers. Here, a worsening of the situation for important connections cannot be compensated by a better synchronization for unimportant connections. In addition, the importance of a line can change during the day. In the morning, the rush hour leads into the city, but in the evening the opposing direction from the city to the suburbs is more important. In our model of the TTSP, we consider the passenger numbers as weights for the corresponding transfers.

The passenger numbers are a good measure of how important a transfer is. An easy approach would be to weight the transfer penalties according to the number of passengers that use this transfer. But since these numbers are not available on the level of detail that is needed, we use a different concept. For our approach, the traffic planner evaluates the situation on different levels. These levels are

- the network nodes,

- the connection types (arriving line cluster, departing line cluster, network node),

- the connections (arriving line (+ direction), departing line (+ direction), network node).

For each element of these levels, a rank of the type "A", "B", "C" or "I" is chosen. "A" means that the regarded object is important compared to the other ones of this level, "B" refers to a normal situation and "C" denotes a situation that is comparatively minor important for the given network. The rank "I" refers to an "inactive" object that is not regarded for the optimization.

**Definition 2.18 (ABC Classification)** *An ABC classification is a mapping from all network nodes, connection types and connections to the set* $\{A, B, C, I\}$.

These classifications can influence the optimization approaches in two ways. In a penalty-based approach, each transfer gets a multiplier for its penalty value that depends on its ABC classification. Here, we take for an "A" the factor 5, for a "B" the factor 3 and for a "C" the factor 1. Instead of ignoring the objects that are classified "I", one can also set a factor of 0 here. Since each transfer belongs to a unique network node, a unique connection type and a unique connection, it gets an overall factor that is the product of three ABC classification factors.

A different approach is to optimize the timetable so that many important connections (rank "A") are improved and only a few of the important connections are worsened. This approach has the advantage that the results are easily understandable and that they do not contain abstract penalty numbers. We introduce the optimization goals and objective functions in detail in Section 2.5.

## 2.4.4   Changing the System

Timetables are grown structures whose interconnectedness makes it quite hard to change the given system without destroying important connections that the passengers are accustomed to. In addition, the vehicle scheduling of the companies and some periodic aspects of the timetable design restrict the possibilities for changes a lot. Therefore, our approach allows only changes to the starting times of whole lines. These changes have two main advantages. Firstly, they do not destroy the periodic structures that are a very important aspect of the timetables, especially in urban areas. And secondly, most vehicle schedules are not affected by these changes, since a bus often serves only a single line.

The following definition introduces these possible changes for the starting times of lines.

**Definition 2.19 (Shift)** *A shift $s$ of a line $l$ defines a change of the starting times of all tours $\zeta \in l$. Thus, for each station $\omega$ that is part of $\zeta$, the arriving time $t_{arr}(\omega)$ and the departing time $t_{dep}(\omega)$ are shifted by $s$ minutes. Since timetables are planned in integral times, we only allow shifts $s \in \mathbb{Z}$. For a line $l_i$, we define the set of allowed shifts by $S_i = \{s_{i1}, ..., s_{in_i}\}$ with $s_{ij} \in \mathbb{Z}$. In addition, $n_i$ represents the number of allowed shifts for line $l_i$. If a shift for a line $l_i$ is chosen, we will refer to it as $s_i \in S_i$.*

**Remark 2.20** *The mathematical concepts that we introduce later in this thesis also work with an approach of shifts for tours. But the combinatorial multiplicity of combinations of shifts would not allow the given algorithms to find good solutions within a reasonable time. The problem can be formulated as a Quadratic Semi-Assignment Problem, which is known to be NP-hard. Thus, finding optimal or near-optimal solutions is very difficult for bigger problem sizes.*

Assigning good shifts to lines is one of the main issues of this work. The problem is to assign to each line a corresponding shift so that a good synchronization is reached. Such an assignment is defined as follows.

**Definition 2.21 (Assignment)** *An assignment $A$ is a function*

$$A : L \to \mathbb{Z}, \quad A : l_i \mapsto s_i$$

*that maps all lines to shifts. Each line $l_i \in L$ is mapped to a feasible shift $s_i \in S_i$. The set of all feasible assignments is denoted by $\mathcal{A}$.*

## 2.4.5   Input data

The presented concept is very dependent on the input data. For an adequate model, the following information must be predefined:

- **stations**: the set of important stations where passengers often transfer must be chosen from the list of all stations in a network,

- **network nodes**: the stations that are chosen must be partitioned into network nodes. Here, stations that are in walking distance and that propose good transfer possibilities may be combined to a network node,

- **lines**: important lines, whose transfer quality should be improved, must be chosen from the list of all lines,

- **line clusters**: the lines that are chosen must be partitioned into line clusters. An example is to group the lines from the same operator that share similar characteristics to a line cluster,

- **walking time matrices**: the walking time matrices for all network nodes must be filled. Each entry represents the walking distance (in minutes) from a pair (arriving station, arriving line cluster) to a pair (departing station, departing line cluster),

- **time intervals**: the time interval of the analysis must be defined (e.g., Saturday morning, from 10 a.m. to 2 p.m.),

- **waiting time classifications**: for each combination of (network node, arriving line cluster, departing line cluster), the time intervals for almost transfers, risk transfers, convenience transfers and patience transfers must be defined,

- **penalty functions**: for each waiting time classification, the penalty function $P$ must be defined. It is then automatically transformed into the smoothed penalty function $\widehat{P}$,

- **ABC classifications**: the ABC classification can be made to influence the optimization. Alternatively, all objects are regarded as "B".

## 2.5   Goals and Objective Functions

In this section, we introduce the goals of the traffic planners. These goals are constructed either as objective functions for the optimization process to reflect the priorities of the planners or as constraints that guide the optimization to find solutions that are acceptable in terms of solution quality.

### 2.5.1   Multi-Criteria Aspects of the Problem

One interesting aspect of the TTSP that makes it such a hard task to solve is the multi-criteria structure of the problem. Every single transfer is in a wider sense an optimization criteria on its own. In other words, it is not sufficient to optimize only the average quality of all transfers, since even if this average can be improved significantly, there would be too many transfers that get lost.

The fact that an improvement of all transfers is not possible can be illustrated by the simple model of two tours that meet at a station, cf. Figure 2.5. Both transfers are opposing, which means that shortening the waiting time of the first transfer results in most cases in a lengthened waiting time for the second transfer.

Figure 2.5: Example of two tours that generate two opposed transfers. The long arrows represent tours and the short arrows refer to transfer possibilities.

If we extend this example to two meeting lines instead of tours, there are already eight opposing transfer possibilities, cf. Figure 2.6. Even in this small example, the high complexity of the model can be observed.

Each connection in this example has an opposing connection and while trying to shorten the waiting times of the transfers of one connection, the waiting times of the transfers of the opposing connection get lengthened.

If a whole network is considered, one can easily see that the system has a complicated interconnectedness which is impossible to consider completely

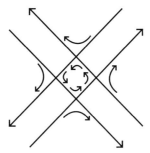

Figure 2.6: Example of two lines that generate eight connections. Both lines have tours in opposing directions (forward, backward).

in a manual planning process. Thus, because of this high interconnectivity, it is not possible to regard one transfer after the other and improve them individually. Instead, the whole network must be considered for every change that is made.

## 2.5.2 Global vs. Local Optimization

In addition to the multi-criteria structure, it is important to differ between a global and a local view on the given problem. The global aspects mean that the traffic planners try to improve the whole network but are confronted with the huge amount of data and the problems arising from the interconnected-ness of the system. The local aspects include the detailed information of the given transfers. Is a bus important for pupils to reach their school? Is a bus-train transfer important for the rush hour and must be preserved? Is the departing time of a train unchangeable due to the railway network? Those are just some of the examples that influence the possibilities for changes to the actual timetable.

For a better understanding of the contrasts between the local and the global view, we introduce five different levels of detail. These levels that help to analyze the timetables both globally and locally are presented in Figure 2.7.

All these levels of detail have different aspects that the traffic planners want to observe:

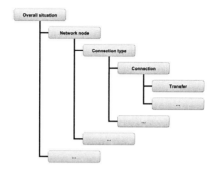

Figure 2.7: Illustration of the five different levels of detail of our concept.

- Overall situation: the global overview of the network,

- Network node: the situation at an important transfer location (e.g., the main station),

- Connection type: all transfers between two companies at a network node (e.g., the connections between the buses and trains at a station),

- Connection: the situation between two lines (with directions) at a network node,

- Transfer: the actual transfer relations for the passengers.

Given these levels of detail, a good approach is to combine state-of-the-art optimization techniques and the expert knowledge of the traffic planners. With modern computers, the high amount of transfers can be observed simultaneously, and so the global problem to improve the whole network can be solved. Additionally, the traffic planner can add his special knowledge about the local situation to help the program to find feasible solutions.

This approach can be implemented in an interactive optimization process, in which the mathematical algorithm presents good solutions that the traffic planner analyzes. Afterwards, he can influence the next iteration of the algorithm with new constraints so that the next computation step can generate solutions that fit his needs better. This process is repeated until a solution is found that satisfies the requirements. The interactive process is visualized in Figure 2.8.

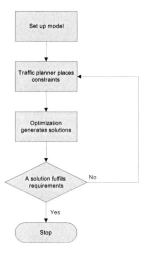

Figure 2.8: Illustration of the interactive optimization process.

## 2.5.3   Pareto Optimality

While trying to find global optimal solutions of multi-criteria problems, the difficulty occurs that it is not easy to compare different solutions.

One way to deal with this problem is to weight the objective functions and to combine them as a sum into a new single objective function. This approach has the disadvantage that it is often not possible to find the correct weights that correspond to the goals of the traffic planner.

A different approach is to find solutions that are called Pareto optimal. These solutions have in common that none of them is dominated by any other solution. A solution dominates another solution, if it is not worse in any objective function and better in at least one objective function. Thus, for a given Pareto optimal solution, one can only improve one objective function by worsening another one.

The set of all Pareto optimal solutions of a problem instance is called Pareto frontier. The goal of the optimization process is now to approximate this Pareto frontier efficiently. Different heuristics that generate such an approximation for the TTSP are presented in Chapter 4.

## 2.5.4  Two Approaches for Objective Functions

In this subsection, we present two different approaches of how to define the objective functions for the TTSP. The difference between these approaches is that the first one has a more analytical character while the second one is constructed so that it is easily understandable by traffic planners and that it is independent of the decisions that must be made for the penalty functions. The computational results of the different approaches are presented in Section 7.3.

**First approach, numerical analysis**

For the first approach, we define three objective functions $\psi_1$, $\psi_2$ and $\psi_3$ that model different goals of the traffic planner.

The first objective function $\psi_1$ describes the overall quality of a timetable. The penalties of all transfers are summed up to get a measure of the average transfer quality in a network.

Let $A \in \mathcal{A}$ be an assignment that chooses for every line $l_i$ a corresponding shift $s_i \in S_i$. The function $\xi(A, c, \zeta_{arr})$ describes the waiting time for the arriving tour $\zeta_{arr}$ at connection $c$ that results from assignment $A$. The function $\hat{P}_c(w)$ returns the penalty value at a connection $c$ for a specific waiting time $w$. With the set $T_{arr}^c$ of all tours of the arriving line $l_{arr}$ for the connection $c$, we get the first objective function by

$$\psi_1(A) = \sum_{c \in C} \sum_{\zeta \in T_{arr}^c} \hat{P}_c(\xi(A, c, \zeta)).$$

The problem of adding up all penalties is that winners and losers of the new timetable interfere with each other. Thus, it is not possible to tell by the overall penalty $\psi_1$, whether the underlying timetable is a good solution. For example, big improvements at one network node can make worsenings at another network node unrecognizable.

The second objective function solves this problem. Here, we take the average transfer quality for every network node into account. The objective function $\psi_2$ describes the highest percentage loss at a single network node. Thus, by minimizing $\psi_2$, a big worsening of the local situation at a single network node can be avoided.

For a network node $\delta \in \Delta$, let $C_\delta$ be the set of all connections that belong to $\delta$. Let $A_0$ be the assignment that sets every shift to zero minutes ($s_i = 0$

for all $i \in M$). Thus, it represents the current timetable situation. The goal of the second objective function is to minimize

$$\psi_2(A) = \max_{\delta \in \Delta} \left( \frac{\sum_{c \in C_\delta} \sum_{\zeta \in T_{arr}^c} \hat{P}_c(\xi(A, c, \zeta))}{\sum_{c \in C_\delta} \sum_{\zeta \in T_{arr}^c} \hat{P}_c(\xi(A_0, c, \zeta))} \right).$$

It is easy to see that $\psi_1$ and $\psi_2$ are correlated. Since $\psi_2$ refers to the worst situation at a network node, the percentage loss (or gain) of the whole network cannot be worse than the percentage loss (or gain) at this worst network node. A visualization of the correlation is presented in Chapter 7.

By minimizing both objective functions $\psi_1$ and $\psi_2$, timetables are created that have a good overall quality and that have no big losses at single network nodes. But the approach of this work is to synchronize existing timetables by making small changes and not to create totally new timetables. The third objective function $\psi_3$ that is closely related to the variance from probability theory deals with this kind of problem.

Let $A(l_i)$ be the shift $s_i$ that is assigned to line $l_i$ by assignment $A$. For a specific assignment $A$ and a line cluster $\lambda$, the average shift value $\vartheta(A, \lambda)$ of assignment $A$ for line cluster $\lambda$ is defined by

$$\vartheta(A, \lambda) = \frac{\sum_{l \in \lambda} A(l)}{\sum_{l \in \lambda} 1}.$$

With the set of all line clusters $\Lambda$, we get the third objective function by

$$\psi_3(A) = \sum_{\lambda \in \Lambda} \sum_{l \in \lambda} (A(l) - \vartheta(A, \lambda))^2.$$

By combining $\psi_1$, $\psi_2$ and $\psi_3$, we are able to generate solutions that fulfill the needs of the traffic planners.

For a better understanding of the structure of the different objective functions $\psi_1$, $\psi_2$ and $\psi_3$, we present some three dimensional visualizations. The first two visualizations shown in Figure 2.9 present the objective functions $\psi_1$ and $\psi_2$ with the possible shifts for two lines. A smoothed penalty function that is not rounded is taken into account. Thus, not only integral values are possible penalties in this example. In addition, shifts are regarded as continuous variables for these visualizations. One can see the high amount of local minima that occur even if only two lines are taken into account. Note that the surface of the second objective function $\psi_2$ is truncated in some areas.

This situation occurs, if the lines are shifted so that the network node at which the worst situation occurs changes and the visualized lines do not stop at this new worst network node. In Figure 2.10, the convex surface of $\psi_3$ is presented for two lines of the same line cluster (left) and for two lines of different line clusters (right).

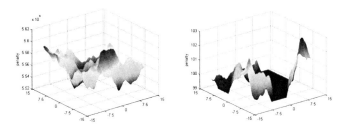

Figure 2.9: 3-Dimensional illustration of $\psi_1$ (left) and $\psi_2$ (right). Two lines are regarded that can both be shifted by $-15$ minutes to $+15$ minutes.

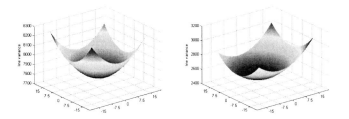

Figure 2.10: 3-Dimensional illustration of $\psi_3$ with lines from the same line cluster (left) and with lines from different line clusters (right). Two lines are regarded that can both be shifted by $-15$ minutes to $+15$ minutes.

**Second approach, intuitive analysis**

A second and more intuitive approach to define objective functions for the TTSP is to directly compare the number of gained transfers versus the number of lost transfers. This evaluation can be refined by counting only those transfers that belong to an important connection (rank "A" in the ABC classification).

This approach has the advantage that a solution that says '50 gained "A" transfers' versus '10 lost "A" transfers' is easily understandable. Nevertheless, the traffic planner still has to analyze the solutions for unwanted worsenings.

With this approach, several objective functions can be constructed easily. For each of these functions, we need the transfer type changes (old transfer types to new transfer types) that should be regarded, an ABC categorization on the ranked levels and whether the number of transfers that fulfill these requirements should be maximized or minimized. Two examples of some chosen transfer type changes are shown in Table 2.2 and in Table 2.3. Here, all newly generated "real" transfers are counted. If combined with the ABC categorization that all transfers must have connection rank "A" and the transfer types and network nodes may have rank "A" or "B", we have easily generated an objective function that should be maximized. Results for this approach are presented in detail in Section 7.3.

| from \ to | almost | risk | conv. | patience | no tr. |
|---|---|---|---|---|---|
| almost | | + | + | + | |
| risk | | | | | |
| convenience | | | | | |
| patience | | | | | |
| no transfer | | + | + | + | |

Table 2.2: Example of a transfer type change table that is used to generate an objective function. The goal is to maximize the number of chosen transfer type changes (gained real transfers).

## 2.6 Periodic Planning and Vehicle Scheduling

A big problem that occurs while synchronizing timetables is that timetables are grown structures. The complicated network with the high amount of vehicles cannot be changed without paying attention to important facts like the

| from \ to | almost | risk | conv. | patience | no tr. |
|---|---|---|---|---|---|
| almost | | | | | |
| risk | + | | | | + |
| convenience | + | + | | + | + |
| patience | + | | | | + |
| no transfer | | | | | |

Table 2.3: Example of a transfer type change table that is used to generate an objective function. The goal is to minimize the number of chosen transfer type changes (lost real transfers and lost convenience transfers).

periodic structure of the timetables and the underlying vehicle schedule. An example of a periodic timetable planning is given by Liebchen in [35]. Here, the Berlin underground network is optimized with mathematical methods.

## 2.6.1   Periodic Planning

In urban areas, most timetables are based on a periodic schedule with a period that is a divisor or a multiple of one hour. This means that most tours of a periodic line start with constant time distances. This structure has the advantage that it is good to remember for the passengers and that the synchronization of the lines repeats every hour. Thus, it is much easier for the traffic planner to achieve a good synchronization. We define a period in this work as follows.

**Definition 2.22 (Period)** *A period of a line $l$ for a specific time interval $[t_{start}, t_{end}]$ and a start station $\omega$ is the smallest time $p$ for which there is every $p$ minutes a departure of a tour of line $l$ at station $\omega$ beginning with a starting time $t_0 \in [t_{start}, t_{start} + p]$.*

According to this definition, a line must not have the same period for the whole day or for the whole set of stations that it serves. It is also possible that a line has a certain period only at a section of its track. This is the case if two different routes are combined in one line that serve the same stations in the central station parts but that serve different stations at the beginning and the end of their routes.

For periodic timetables, it is often not feasible to change the starting times of single tours, since such a change would destroy the periodic structure of the timetable. An approach to avoid the destruction of the periods is to shift only whole lines. This approach was applied to the bus network of the city of Kaiserslautern in 2004. Here, several bus lines were shifted by the same amount of time to generate a better synchronization between the buses and the trains at the main station. For this optimization only the two shift possibilities $-7$ and $+8$ minutes were considered, since the outcome was analyzed manually.

## 2.6.2 Vehicle Scheduling

In contrast to the periodic timetable structure in urban areas, it is often not economical to have a periodic timetable with a high density in rural areas. Here, the vehicle scheduling plays a much bigger role, meaning that the amount of buses and the availability of drivers influences the times and routes of the buses much stronger. Shifting a tour often influences the vehicle schedule of the corresponding bus and the traffic planner has to keep in mind that the vehicle schedule must be feasible for the generated new timetables.

An example of an interconnected vehicle scheduling can be seen in Figure 2.11. Shifting the starting time of line 1 can cause a problem when the same bus will serve line 2 later, since the arrival of the tour of line 1 must be before the planned departure of the tour of line 2.

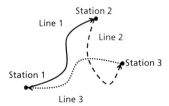

Figure 2.11: Example of a vehicle scheduling circle with three lines. The vehicle starts by driving a tour of line 1, followed by tours of line 2 and line 3. Afterwards, the circle starts again with a tour of line 1.

The main timetable synchronization ideas of our approach are combined with vehicle schedule preservation techniques by Hansen in [26]. In this work, the problem of finding cost optimal vehicle schedules in variable timetables is analyzed. Here, the goal is to minimize the sum of the penalty values and the costs of serving all tours in a multi-criteria approach. One conclusion of this work is that purely minimizing the penalty will very likely result in an increasing fleet size, while maintaining or reducing the fleet size does not rule out an improvement of the transfer situation.

## 2.7   Constraints

In addition to the objective functions, different kinds of constraints help the traffic planner to integrate his expert knowledge into the model. But unlike the objective functions, the constraints do not determine global goals. Instead, they are an approach to consider special cases of real life situations in our model.

This section introduces three types of constraints that refer to different use cases. Additionally, we explain how we treat these constraints in our model.

### 2.7.1   Lineshift Constraints

A lineshift constraint is the most simple type of constraint in our model. It determines a reduction of the possible changes that can be made to the starting times of a line. This type of constraint is required if a line should not be shifted backwards, for example, if a school bus is not allowed to leave before the school ends. Another example of a situation in which a lineshift constraint can be helpful, is a railway network that forbids certain types of shifts, because the rails are already scheduled. The mathematical definition of a lineshift constraint is as follows.

**Definition 2.23 (Lineshift Constraint)** *Given a line $l$, a lineshift constraint defines a reduction of the set of allowed shifts $S_i = \{s_{i1}, ..., s_{in}\}$ to a reduced set $S' \subseteq S$. This new set $S'$ is then used instead of $S$ to determine the search space.*

Especially for train lines, the lineshift constraints play an important role. In most cases, trains connect several cities and a change of the starting time of a train by more than a few minutes in one city will most likely destroy the synchronization in the other cities that this train serves. Thus, allowing only very small shifts for some of the train lines is a feasible approach to deal with this problem.

## 2.7.2 Period Constraints

In Section 2.6, the concept of periodic lines was introduced. In some cities, several periodic lines share a common meeting point that should be maintained in new synchronized timetables. Such a meeting point means that at a specific time, several tours arrive at a station, then all vehicles wait a few minutes to give the passengers the possibility to transfer between the vehicles and afterwards all tours leave the station simultaneously.

**Definition 2.24 (Common Meeting Point)** *For a set of lines $l_1, ..., l_k$ and a station $\omega$, a common meeting point is defined by a time $t$ at which all these lines have a tour that stops at this station.*

If such a meeting point occurs periodically, the situation is very convenient for the passengers, since they have good transfer possibilities and there are not so many departure times that they must keep in mind.

If a network has such a periodic meeting point, the traffic planner wants to maintain it with a high priority. Thus, an assignment to the starting times of these already synchronized lines must preserve the common meeting point at the given station. For a formulation of a constraint that can deal with this problem, we need some number-theoretic basic knowledge.

Suppose that the given lines $l_1, ..., l_k$ have the periods $p_1, ..., p_k$ that are all a divisor of one hour and that there is a common meeting point every $\bar{t}$ minutes. In order to keep these lines synchronized at a particular station, it suffices to ensure that they still have a common meeting point every $\bar{t}$ minutes. Note that the meeting times and the tours that actually meet can be different from the ones of the current timetable, since the meeting points regard whole lines and not single tours. An example of such a synchronization can be seen in Figure 2.12, where three lines with periods of 15, 20 and 30 minutes are regarded and 10 a.m. and 11 a.m. are such meeting points.

The definition of a constraint that preserves these meeting points is as follows.

**Definition 2.25 (Period Constraint)** *Given are a set of lines $l_1, ..., l_k$ with*

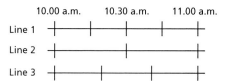

Figure 2.12: Example of three synchronized lines at a station. The vertical bars represent arrivals of tours of the different lines. The lines have periods of 15, 20 and 30 minutes.

*periods $p_1, ..., p_k$ that have a common meeting point. A period constraint enforces these lines to keep all common meeting points if shifts are applied.*

We present a constructive way to find feasible shift combinations that will not destroy the common meeting points of the given lines. It will not destroy the synchronization, if all synchronized lines are shifted by the same time $t_0$. Furthermore, a line can be shifted by its period without affecting the transfer possibilities. Thus, we get for all $\nu_i \in \mathbb{Z}$ the feasible shifts

$$s_i = t_0 + \nu_i \cdot p_i \quad \forall i \in \{1, ..., k\}$$

or in a different notation

$$s_i \equiv t_0 \pmod{p_i} \quad \forall i \in \{1, ..., k\}.$$

But there are even more changes that are allowed for these lines. For a shorter notation, we introduce

$$z_{i,j} := \gcd(p_i, p_j)$$

as the greatest common divisor of the periods of line $l_i$ and line $l_j$. Furthermore, let

$$z_i := \operatorname{lcm}(z_{i,1}, ..., z_{i,i-1}, z_{i,i+1}, ..., z_{i,k})$$

be the least common multiple of all pairs of greatest common divisors of the period $p_i$ and the periods $p_j$ ($j \in \{1, ..., k\} \setminus \{i\}$) of all other lines.

For our further analysis, we need the Generalized Chinese Remainder Theorem. It is about the solvability of simultaneous congruences.

**Theorem 2.26 (Generalized Chinese Remainder Theorem)** *For a set $\{n_1, ..., n_k\}$ of positive integers that are pairwise coprime and any numbers $i_1, ..., i_k \in \mathbb{Z}$, there exists an integer solution $x$ that solves the system of simultaneous congruences*

$$x \equiv i_1 \pmod{n_1}$$
$$x \equiv i_2 \pmod{n_2}$$
$$\vdots$$
$$x \equiv i_k \pmod{n_k}.$$

*All solutions $x$ to this system are congruent modulo $N = n_1 \cdot ... \cdot n_k$. Thus, for two solutions $x$ and $y$, we have*

$$x \equiv y \pmod{n_i} \quad \forall i \in \{1, ..., k\} \text{ if and only if } x \equiv y \pmod{N}.$$

*Furthermore, if the integers $n_1, ..., n_k$ are not pairwise coprime, a solution exists if and only if*

$$i_u \equiv i_v \pmod{\gcd(n_u, n_v)} \quad \forall u, v \in \{1, ..., k\}.$$

*In this case, all solutions $x$ to this system are pairwise congruent modulo the least common multiple $lcm(n_1, ..., n_k)$.*

With this theorem, we can now prove that shifting the given lines additionally by the values $z_i$ will result in a timetable that still has a common meeting point of the synchronized lines every hour.

**Theorem 2.27** *If the periodic lines $l_1, ..., l_k$ with a common meeting point and periods $p_1, ..., p_k$ that are divisors of a time interval of $p$ minutes are shifted by*

$$s_i = t_0 + \nu_i \cdot p_i + \mu_i \cdot z_i \quad \forall i \in \{1, ..., k\}$$

*with $\nu_i, \mu_i \in \mathbb{Z}$, there will be at least one common meeting point every $p$ minutes.*

**Proof:**

To prove the theorem, we will construct for each common meeting point $m_p$ the new shifted meeting point $m'_p$. With the given shifts

$$s_i = t_0 + \nu_i \cdot p_i + \mu_i \cdot z_i \quad \forall i \in \{1, ..., k\},$$

we get the equations

$$x' = t_0 + \nu'_1 \cdot p_1 + \mu'_1 \cdot z_1,$$

$$...$$

$$x' = t_0 + \nu'_k \cdot p_k + \mu'_k \cdot z_k,$$

where $x'$ is the time distance between $m'_p$ and $m_p$ ($\nu'_i, \mu'_i \in \mathbb{Z}$). We transform these equations into the simultaneous congruences

$$x \equiv \mu_1 \cdot z_1 \pmod{p_1},$$

$$...$$

$$x \equiv \mu_k \cdot z_k \pmod{p_k}$$

with $x = x' - t_0$.

We can apply the Generalized Chinese Remainder Theorem (Theorem 2.26) to this congruence system. Note that the periods $p_1, ..., p_k$ are not necessarily coprime. These simultaneous congruences have a solution if and only if

$$\mu_i \cdot z_i \equiv \mu_j \cdot z_j \pmod{z_{i,j}} \quad \forall i, j \in \{1, ..., k\}.$$

We have per definition that $z_{i,j}$ divides $z_i$. Hence, this condition is fulfilled and the system has a solution $x$. Furthermore, we know that each $x^* = x + \nu \cdot \operatorname{lcm}(p_1, ..., p_k)$ is a solution of the system and that $\operatorname{lcm}(p_1, ..., p_k) \leq p$ holds.

The existence of a solution of the given congruence system proves that all common meeting points are preserved if only the corresponding shifts are applied. Each old meeting point is shifted by $x' = x + t_0$ minutes. The solution $x$ is generated modulo $\operatorname{lcm}(p_1, ..., p_k)$. Thus, there exists a solution every $\operatorname{lcm}(p_1, ..., p_k)$ minutes which proves the theorem.

$\square$

**Remark 2.28** *We know that $z_{i,j}$ divides $p_i$ for all $j \in \{1, ..., k\}$, since $z_{i,j} = gcd(p_i, p_j)$. Thus, we also have $z_i | p_i$ and we can transform the formulations from Theorem 2.27 into*

$$s_i = t_0 + \mu_i \cdot z_i \ or$$

$$s_i \equiv t_0 \pmod{z_i}.$$

*This means that we can shift each line $l_i$ by multiples of the corresponding value $z_i$ without destroying its common meeting points.*

### 2.7.3 Connection Constraints

The lineshift constraints and the period constraints refer to global situations that regard lines and tours. But there is still a lack of control for the single transfers. Therefore, we introduce the concept of connection constraints.

With the help of these constraints, the traffic planner can influence our model in a very detailed way. Here, the allowed transfer types for a given transfer can be restricted. He can, for example, define that a convenience transfer must be preserved or that an almost transfer must be transformed into a real transfer, i.e., a risk transfer, a convenience transfer or a patience transfer. These connection constraints present an intuitive way for the traffic planner to influence the model, without being forced to directly access the mathematical formulation.

**Definition 2.29 (Connection Constraint)** *A connection constraint specifies a set of allowed transfer types for one or several transfers that belong to the same connection. Thus, for a connection $c$ and the set of all transfers $\mathcal{T}_c$ of this connection, a connection constraint is defined by a mapping from a subset of the transfers $T' \subseteq \mathcal{T}_c$ to a subset of the transfer types {no transfer, almost transfer, risk transfer, convenience transfer, patience transfer}.*

A connection constraint can be transformed into a disjunction of conjunctions of allowed shifts. This means that for the shift difference

$$s_{diff} = s_{arr} - s_{dep}$$

of the shift of the arriving line $s_{arr}$ and the shift of the departing line $s_{dep}$, only some intervals are allowed (e.g., $s_{diff} \in \{i_1, ..., i_2\} \cup \{i_3, ..., i_4\} \cup ...$ with $i_1 < i_2 < ...$). Such a disjunction can be seen in the following example that helps to get a better understanding of the concept.

**Example 2.30** *Given are an arriving tour $\zeta_{arr}$ of line $l_{arr}$ that arrives at 3 p.m. at a network node that consists of a single station. We analyze the transfer to the departing line $l_{dep}$ that has three departing tours with the correct direction in a relevant time interval. These tours $\zeta_{dep_1}, \zeta_{dep_2}$ and $\zeta_{dep_3}$ leave the station at 2.48 p.m., at 3.08 p.m. and at 3.28 p.m. in the current timetable. The allowed shifts for both lines are $\{-10, ..., +10\}$ minutes and we get $s_{diff} \in \{-20, ..., +20\}$.*

Let the convenience time interval be from 3 to 10 minutes and assume a walking time inside this network node of zero minutes. The transfer of the current timetable has a waiting time of 8 minutes, since $\zeta_{dep_2}$ is the best transfer possibility for the passenger. Thus, the transfer is a convenience transfer.

The traffic planner wants to preserve this convenience transfer and defines a connection constraint. For each of the departing tours, various shift differences lead to a convenience transfer. These allowed shift differences are for

- tour $\zeta_{dep_1}$: $I_1 = \{-20, ..., -15\}$,

- tour $\zeta_{dep_2}$: $I_2 = \{-2, ..., 5\}$,

- tour $\zeta_{dep_3}$: $I_3 = \{+18, ..., +20\}$,

and we get a convenience transfer, if

$$s_{diff} \in I_1 \cup I_2 \cup I_3.$$

While generating new timetables, only those timetables that fulfill this constraint are feasible.

### 2.7.4   Binary Formulation of the Constraints

For a further application of the constraints in the mathematical formulation of the problem, we need the constraints to be formulated binary. This means that each constraint can be formulated as a union of formulations that depend maximal on two different lines. The following lemma shows that the three introduced constraints can be formulated as binary constraints.

**Lemma 2.31** *Lineshift constraints, connection constraints and period constraints can be formulated by a union of formulations, where every single formulation only depends on two lines.*

**Proof:**

A line shift constraint reduces the possible shifts for a given line. Here, only the line whose shift possibilities are reduced is needed to formulate the given constraint.

A connection constraint refers to the transfers between two lines $l_i$ and $l_j$. Hence, it is obviously a binary constraint. Each connection constraint can be formulated as a disjunctive normal form (DNF).

For a binary formulation of the period constraints, we need a formulation of the allowed shifts $s_i$ from Theorem 2.27 that is a union of formulations that only contain two different lines. In the current form, the variables $\nu_i$ and $\mu_i$ can still depend on the other lines.

For this new formulation, we generate again a new meeting point with the help of the chosen shifts and the given periods of the lines. This new meeting point can be constructed by

$$x = s_1 + \mu_1 \cdot p_1 \quad \Rightarrow x \equiv s_1 \pmod{p_1},$$

$$\cdots \qquad\qquad \cdots$$

$$x = s_n + \mu_n \cdot p_k \quad \Rightarrow x \equiv s_k \pmod{p_k}.$$

This simultaneous congruence has a solution modulo $\mathrm{lcm}(p_1, ..., p_k)$ if and only if

$$s_i \equiv s_j \pmod{\gcd(p_i, p_j)}$$

holds for all $i, j \in \{1, ..., k\}$. As a binary formulation for our problem, this is equivalent to

$$s_i - s_j = \mu \cdot z_{i,j}$$

with an integer variable $\mu$.

With this consideration, we directly get that the period constraint for $k$ lines with periods $p_1, ..., p_k$ can be transformed into a set of modulo constraints

$$s_i \equiv s_j \pmod{z_{i,j}} \quad \forall i, j \in \{1, ..., k\}.$$

As a binary formulation, we get equations of the type

$$s_i - s_j = \mu \cdot z_{i,j}$$

with a not fixed variable $\mu$.

$\square$

## 2.7.5 Hard Constraints vs. Soft Constraints

There are in general two main ways to deal with constraints in optimization approaches. The first one is to treat them as hard constraints, meaning that no constraint is allowed to be violated. This approach has the advantage that all feasible solutions fulfill all requirements that were preset. But this approach has the disadvantage that one probably loses really good solutions that maybe violate only one constraint that is not so important. Or even worse, the problem gets overconstrained and no feasible solution exists at all.

The second approach is to treat the constraints as soft constraints. Here, we allow some constraints to be violated if the improved solution quality compensates this violation. This treatment of constraints has the advantage that the risk to lose high quality solutions is reduced. But the price for this improvement is that each created solution must be analyzed, whether the violation of the constraints is acceptable. In addition, a comparison of two solutions that violate different kinds of constraints is not an easy task.

In our approach for the TTSP, we treat the constraints as hard constraints. Nevertheless, by some circumstances the problem can get overconstrained or such a high amount of constraints can be set that the heuristics are not able to find any feasible solution at all. In such a case, the hard constraint approach terminates and the optimization algorithm tries to find solutions that violate the least possible number of constraints.

## 2.8 Summary of the Chapter

In this chapter, we introduce a complex model for the TTSP. Some simplifications, for example the focusing on network nodes and the substitution of passenger numbers by ABC classifications, are made to improve the efficiency of the model. We develop an interactive process, in which the traffic planner and the mathematical optimization methods cooperate to generate good timetables.

We analyze the public transport network on different levels of detail. A global level describes the network as a whole, while the local levels include detailed information about the single transfers. The traffic planner integrates his needs and wishes by setting different kinds of constraints. These constraints can be used to maintain the status quo or to force the algorithms to improve certain transfer possibilities.

To evaluate the timetables, we propose two different objective function approaches. These goals are used in our approach to meet the multi-criteria requirements of the problem.

# Chapter 3

# The Quadratic Semi-Assignment Problem

A common problem in operations research is the problem to assign objects to locations. The Quadratic Semi-Assignment Problem (QSAP) is a special case of such a problem, where the objective function contains quadratic terms. Unlike the widely studied Quadratic Assignment Problem (QAP) that creates bijective assignments, several objects may here be assigned to the same location.

The QAP is a special case of the QSAP for which constraints can forbid certain assignment combinations. Both problems are known to be among the hardest NP-hard problems, meaning that even for relatively small problem sizes (assigning 30 objects to 30 locations) it is very hard to solve them exactly in an acceptable amount of time. This is due to the quadratic terms in the objective function that make it quite hard to find optimal solutions. Additionally, the determination of good lower bounds is also known to be a hard task.

QSAPs are an open field of research. They are, unlike the QAP, not very well studied. This thesis deals with a generalized form of the QSAP that allows different feasible sets of locations for the objects. The theoretical aspects of this type of QSAP were, to the best of our knowledge, not studied before.

We introduce in this chapter our generalized version of the QSAP that we use for the TTSP. Additionally, other interesting applications of the problem are presented. We give an interpretation of the model on graphs and we discuss the currently best known lower bound strategies for the QAP and their usability for the QSAP. Furthermore, we present a polyhedral analysis of our generalized form of the QSAP. Finally, we introduce in this chapter first lower bound strategies for the QSAP.

## 3.1   Introduction of the QSAP

In literature, the name Quadratic Semi-Assignment Problem refers to different types of problems that are not equivalent. Thus, we start this chapter by formulating the QSAP that we are dealing with in this thesis. Our formulation is slightly more general than the most often used formulation. It is presented in the following definition.

**Definition 3.1 (Quadratic Semi-Assignment Problem)** *Given are a set of objects $O = \{o_1, ..., o_m\}$ and a set $L = \{L_1, ..., L_m\}$ which contains the individual location sets $L_i = \{l_1^i, ..., l_{n_i}^i\}$. These sets $L_i$ must not be disjoint. For a shorter notation, we introduce the index sets $M = \{1, ..., m\}$ and $N_i = \{1, ..., n_i\}$ for all $i \in M$.*

*Furthermore, object-location-cost-functions*

$$f_i : L_i \to \mathbb{R} \quad (i \in M)$$

*and object-location-pair-cost-functions*

$$f_{i,k} : L_i \times L_k \to \mathbb{R} \quad (i, k \in M)$$

*are needed.*

*The QSAP is the task to assign to each object $o_i \in O$ a location $l_j^i \in L_i$ so that the sum of the corresponding object-location-costs and object-location-pair-costs is minimized. Thus, we get an objective function of the form*

$$f(x) = \sum_{i \in M} \sum_{j \in N_i} f_i(l_j^i) x_{ij} + \sum_{\substack{i,p \in M \\ i \neq p}} \sum_{j \in N_i} \sum_{q \in N_p} f_{i,p}(l_j^i, l_q^p) x_{ij} x_{pq},$$

*where $x_{ij} \in \{0, 1\}$ defines, whether object $o_i$ is assigned to location $l_j^i$. To simplify the notation, we reformulate the objective function by replacing the cost functions with constant coefficients. Thus, we get the new formulation*

$$f(x) = \sum_{i \in M} \sum_{j \in N_i} b'_{ij} x_{ij} + \sum_{\substack{i,p \in M \\ i \neq p}} \sum_{j \in N_i} \sum_{q \in N_p} c'_{ijpq} x_{ij} x_{pq}.$$

*If the coefficients $c'_{ijpq}$ have a symmetric structure of the form*

$$c'_{ijpq} = c'_{pqij} \quad \forall j \in N_i, \ \forall q \in N_p,$$

*we can reformulate the objective function to a notation of the form*

$$f(x) = \sum_{i=1}^{m} \sum_{j=1}^{n_i} b_{ij} x_{ij} + \sum_{i=1}^{m-1} \sum_{p=i+1}^{m} \sum_{j=1}^{n_i} \sum_{q=1}^{n_p} c_{ijpq} x_{ij} x_{pq}.$$

The TTSP has such a symmetric structure. Thus, we focus our research in this thesis on this symmetric formulation.

The QSAP is NP-hard. This result is, e.g., shown by Saito et al. in [49], where the QSAP is reduced to the Hub Location Problem (HLP). The HLP is known to be NP-hard, even if there are only three hubs involved (cf. [58]). This means that it is nearly hopeless to find an optimal solution for a larger QSAP instance in a reasonable time.

In addition to the pure formulation of the QSAP, we allow the problem to have constraints that forbid certain object-location-pair combinations. These constraints are motivated by the constraints from Section 2.7.

## 3.2 Literature Survey

Although the QSAP is not so widely studied in literature, there are some quite interesting papers, both on theory and application.

Applications of the QSAP can be found in [4], [19] and [21]. Here, a task allocation problem, an approach to optimize textile engineering and a biochemical question from the area of protein design are formulated as QSAPs. Different solution strategies, each using the special characteristics of the underlying problem, are presented and applied to the given problems.

The QSAP is closely related to the widely studied QAP. A good overview of recent advances for the QAP is presented in [65]. An interesting polyhedral study of the QAP is done in [29] and refined for the symmetric case in [28]. The authors construct the convex hull of the feasible set of solutions of the problem and an affine transformation is used to determine the dimension of the corresponding polytope. Furthermore, different types of cutting planes are introduced and analyzed. The ideas of these works are also applied to the QSAP in [49]. This work is refined in [48] for the special case of the symmetric QSAP.

A more general problem formulation, the Generalized Quadratic Assignment Problem (GQAP), is introduced in [34]. It describes the assignment problem of $m$ objects to $n$ locations, where each location has a fixed capacity. For this type of problem, solution strategies are presented in [23] and in [43].

The QSAP and the QAP are both known to be NP-hard. A polynomially solvable class of QSAPs is presented in [38], but this class presents a special case that is not related to the QSAP that arises from the TTSP. Therefore, lower bound approaches are needed to get at least an estimation of the optimal solution value for bigger QSAP sizes. Similar considerations are made for the QAP. In particular, two strategies provide good lower bounds for this type of problem. The first one is the Reformulation Linearization Technique (RLT) from [56]. Recent improvements were achieved for the QAP in [1] (2007) with an RLT level 2 approach and in [24] (2008) with an RLT level 3 approach. The second strategy that is also able to generate promising bounds for the QAP is a Semidefinite Programming (SDP) approach that is presented in [63] and in [64].

A different method that provides bounds for the QSAP is the reduction approach from [5]. Here, the structure of the objective function is changed to extract a constant term that can be used as a lower bound. This approach shows promising results if combined with other lower bound approaches. Another lower bound approach, which is based on a polynomially solvable class of QSAPs, is presented in [39].

## 3.3   Applications of the QSAP

Assigning objects to locations is just an example of a problem that can be formulated as a QSAP. In this section, we present several real life applications, including the TTSP from Chapter 2.

### 3.3.1   QSAP for the Timetable Synchronization Problem

As explained in the previous chapter, the TTSP consists of assigning shifts to public transport lines. In terms of Definition 3.1, the lines correspond to the objects and the shifts correspond to the locations of the QSAP. Thus, we have a Semi-Assignment Problem, since the same shift can be assigned to several lines.

The quality of a network arises from the transfer situation between pairs of lines. This means that for every pair of lines with chosen shifts, there is a quality measure for the new transfer situation between these lines, depending on the convenience of the passengers. With the objective function $\psi_1$ from Section 2.5, we have a quadratic problem. The objective function contains no linear terms $b_{ij}$, since there are no costs or handicaps that arise from assigning a shift to a line.

The problem sizes depend on the size of the network that is regarded. Smaller networks may have less than ten lines that can be shifted, but for bigger cities, the number of lines can grow up to 50. Constraints that reflect special information about the traffic network forbid certain shifts or shift combinations.

### 3.3.2  Task Allocation

One of the first applications of the QSAP that was studied is the task allocation problem. In 1990, Billionnet et al. presented in [4] an algorithm for a Task Allocation Problem, where tasks are assigned to processors. The approach is a special case of the QSAP. Here, the objective function has a simplified structure that makes the mathematical analysis of the problem easier. The coefficients $b_{ij}$ in the objective function represent the execution costs of task $t_i$ on processor $p_j$ and the coefficients $c_{ijkl}$ represent the communication costs for two tasks $t_i$ and $t_k$, if they are assigned to different processors $p_j$ and $p_l$. The number of communication tasks is low compared to the number of task pairs. This diminishes the objective function, since many coefficients have the value zero.

The authors propose a branch and bound algorithm that uses a Lagrangian Relaxation of the constraints. Zero duality gaps are found by analyzing the consistency of a pseudo-Boolean equation. Problems of size up to 20 processors and 50 tasks or 10 processors and 100 tasks could be solved to optimality.

### 3.3.3   Textile Engineering

The research done in [21] from 2002 deals with the task to route a set of
thread-spools on a machine from their origin to their destination without
collisions. This problem is closely related to our constrained QSAP. A $k$-
partite graph is constructed, where every partition corresponds to a spool
and all vertices inside such a partition define possible routes for this spool.
Two vertices from different partitions are connected by an edge if and only
if the two routes cause no collision of the two spools. If this graph contains
a $k$-clique, a solution that assigns a route to each spool without collisions is
found.

The authors propose a constructive clique generating process that is used
in a branch and bound algorithm. Compared to our problem, the ratio
between the number of partitions and the number of vertices is different.
In textile engineering, the number of routes is very high compared to the
number of spools, since many possible routes are allowed for a spool. The
presented algorithm is an exact approach that can solve small problem sizes.

### 3.3.4   Protein Design

In de novo protein design, amino acid sequences are given as flexible three-
dimensional protein structures. By minimizing the sum of the energy inter-
actions between the amino acid pairs, protein structures with a high stability
and functionality can be designed.

The task is to assign amino acids to positions along the backbone of a
protein. The weights on the edges refer to a detailed atomistic energy force
field. In [19] (2005), the authors compare several mathematical formulations
for the in silico sequence selection problem. One of them is the Reformulation
Linearization Technique that provides good lower bounds for quadratic problems by introducing relaxed linear formulations. This technique is studied in
detail in Chapter 5 in our thesis. Only the first level of RLT approximations
is used in this work. Problems up to a size of 35 amino acids with 20 possible
positions along the backbone are analyzed in this work.

In the context of this work, Pierce and Winfree showed in [44] that the
protein design problem is NP-hard.

## 3.4 MIP-Modeling

The QSAP has the characteristic trait of being easy to formulate but being hard to solve. An Integer Program (IP) formulation of the problem can be given as follows:

$$(IP_0) \quad \min \sum_{i=1}^{m} \sum_{j=1}^{n_i} b'_{ij} x_{ij} + \sum_{i=1}^{m} \sum_{\substack{k=1 \\ k \neq i}}^{m} \sum_{j=1}^{n_i} \sum_{l=1}^{n_k} c'_{ijkl} x_{ij} x_{kl}$$

$$s.t. \sum_{j=1}^{n_i} x_{ij} = 1 \quad \forall i \in M,$$

$$x_{ij} \in \{0, 1\} \quad \forall i \in M, \ \forall j \in N_i.$$

In this work, we deal mostly with the symmetric case, where $c'_{ijkl} = c'_{klij}$ holds. Thus, we simplify the formulation to the following Integer Program:

$$(IP) \quad \min \sum_{i=1}^{m} \sum_{j=1}^{n_i} b_{ij} x_{ij} + \sum_{i=1}^{m-1} \sum_{k=i+1}^{m} \sum_{j=1}^{n_i} \sum_{l=1}^{n_k} c_{ijkl} x_{ij} x_{kl}$$

$$s.t. \sum_{j=1}^{n_i} x_{ij} = 1 \quad \forall i \in M,$$

$$x_{ij} \in \{0, 1\} \quad \forall i \in M, \ \forall j \in N_i.$$

In spite of the very compact formulation, the QSAP withstands all efforts of being solved easily due to the quadratic terms. Thus, we analyze different strategies to find more efficient ways than the Integer Program formulation to deal with this type of problem.

### 3.4.1 MIP Formulation

A first approach to construct a formulation that can be solved by modern solvers is to eliminate the non-linear quadratic terms in the objective function. Here, we replace the products of the type $x_{ij} \cdot x_{kl}$ by new variables $y_{ijkl}$ that must be greater than or equal to zero. With this approach, we get a Mixed Integer Program (MIP) formulation. To maintain the problem structure, a set of constraints is introduced to ensure that these new variables behave like the product of the old variables without using quadratic terms.

These constraints are of the form

$$(C0) \quad \sum_{j=1}^{n_i} x_{ij} = 1 \quad \forall i \in M,$$

$$(C1) \quad \sum_{j=1}^{n_i} y_{ijkl} = x_{kl} \quad \forall i, k \in M \ (i < k), \ \forall l \in N_k,$$

$$(C2) \quad \sum_{l=1}^{n_k} y_{ijkl} = x_{ij} \quad \forall i, k \in M \ (i < k), \ \forall j \in N_i.$$

The already given constraints $(C0)$ of the IP formulation ensure that each object is assigned to exactly one location. The new constraints $(C1)$ and $(C2)$ arise from multiplying $(C0)$ by $x_{kl}$ or $x_{ij}$ and by replacing the emerging product terms by the new variables $y_{ijkl}$. These constraints ensure that the solutions of the MIP correspond to solutions of the IP formulation. To show this, we have to prove that the equation $x_{ij} \cdot x_{kl} = y_{ijkl}$ holds. This is shown in the following proposition.

**Proposition 3.2** *For all $x_{ij}, x_{kl} \in \{0, 1\}$ $(i, k \in M \ (i < k), \ j \in N_i, \ l \in N_k)$ and all $y_{ijkl} \geq 0$ that satisfy $(C0)$, $(C1)$ and $(C2)$, the equation*

$$x_{ij} \cdot x_{kl} = y_{ijkl}$$

*holds. Thus, replacing the product $x_{ij} \cdot x_{kl}$ by the new variable $y_{ijkl}$ does not change the problem structure of the QSAP if the constraints $(C1)$ and $(C2)$ are also included.*

**Proof:**

The newly generated variable $y_{ijkl}$ must have value zero if one of the two variables $x_{ij}$ or $x_{kl}$ equals zero, since the sum of the nonnegative $y$-variables is equal to this $x$-variable (this can be concluded from $(C1)$ and $(C2)$). In addition, if both variables $x_{ij}$ and $x_{kl}$ are equal to one, the corresponding $y$-variable $y_{ijkl}$ must also be equal to one. Otherwise, we would get from $y_{ijkl} < 1$ that there exists another variable $y_{ij'kl}$ with a value greater than zero. This implies $x_{ij'} > 0$ and $x_{ij} + x_{ij'} > 1$ contradicts (C0). Thus, $y_{ijkl} = x_{ij} \cdot x_{kl}$ is proven if $x_{ij}, x_{kl} \in \{0, 1\}$ holds.

$\square$

We can conclude from Proposition 3.2 that the integrality of the $x$-variables enforces the integrality of the $y$-variables. Thus, the IP formulation generates the same solutions as the following MIP formulation:

$$(MIP) \min \sum_{i=1}^{m} \sum_{j=1}^{n_i} b_{ij} x_{ij} + \sum_{i=1}^{m-1} \sum_{k=i+1}^{m} \sum_{j=1}^{n_i} \sum_{l=1}^{n_k} c_{ijkl} y_{ijkl}$$

$$\text{s.t.} \sum_{j=1}^{n_i} x_{ij} = 1 \quad \forall i \in M,$$

$$\sum_{j=1}^{n_i} y_{ijkl} = x_{kl} \quad \forall i, k \in M \ (i < k), \ \forall l \in N_k,$$

$$\sum_{l=1}^{n_k} y_{ijkl} = x_{ij} \quad \forall i, k \in M \ (i < k), \ \forall j \in N_i,$$

$$x_{ij} \in \{0, 1\} \quad \forall i \in M, \ \forall j \in N_i,$$

$$y_{ijkl} \geq 0 \quad \forall i, k \in M \ (i < k), \ \forall j \in N_i, \ \forall l \in N_k.$$

Note that the better solvability of the MIP that is caused by the linearized objective function is achieved for the price of an enlarged problem size. The problem of the complexity of the QSAP is still not solved. We only transferred the non-linear formulation into a linear formulation with a higher number of variables and constraints.

## 3.4.2  LP-Relaxation

By discarding the idea of generating the optimal solution and instead calculating a lower bound, we can reduce the complexity of the formulation significantly. An easy way to do this is to relax the binary condition of the $x$-variables in the MIP formulation. It is sufficient to enforce $x_{ij} \geq 0$, since the $x$-variables cannot become greater than one due to the $(C0)$ constraints.

We transform the given MIP into a Linear Program (LP) formulation.

$$(LP) \ \min \sum_{i=1}^{m} \sum_{j=1}^{n_i} b_{ij} x_{ij} + \sum_{i=1}^{m-1} \sum_{k=i+1}^{m} \sum_{j=1}^{n_i} \sum_{l=1}^{n_k} c_{ijkl} y_{ijkl}$$

$$\text{s.t.} \sum_{j=1}^{n_i} x_{ij} = 1 \quad \forall i \in M,$$

$$\sum_{j=1}^{n_i} y_{ijkl} = x_{kl} \quad \forall i, k \in M \ (i < k), \ \forall l \in N_k,$$

$$\sum_{l=1}^{n_k} y_{ijkl} = x_{ij} \quad \forall i, k \in M \ (i < k), \ \forall j \in N_i,$$

$$x_{ij} \geq 0 \quad \forall i \in M, \ \forall j \in N_i,$$

$$y_{ijkl} \geq 0 \quad \forall i, k \in M \ (i < k), \ \forall j \in N_i, \ \forall l \in N_k.$$

This formulation can be solved by modern LP solvers like Cplex 11.2. The LP formulation generates results much faster than the IP formulation, but the quality of the lower bounds that it generates is unpredictable. The gaps of the solutions are, in most cases, too high to be effectively used in branch and bound algorithms.

A detailed comparison of the results of the LP and the MIP formulation is presented in Section 6.4 of this thesis.

## 3.5   Constraints for the QSAP

In Section 2.7, different constraints for the TTSP are introduced and we have shown that all these constraints can be formulated as binary constraints.

For the QSAP, these constraints can be included as forbidden assignments or forbidden assignment combinations. We now introduce the concept of a constrained QSAP.

**Definition 3.3 (Constrained QSAP)** *A constrained QSAP is a QSAP for which some object-location combinations and/or some object-location-pair combinations are forbidden.*

If several constraints are applied to a QSAP, the question arises whether there is still a feasible solution of the problem. The complexity of deciding the solvability of a constrained QSAP is also NP-complete. This was shown in [44], where the Satisfiability problem was reduced to the constrained QSAP.

We now present a way to formulate the different constraints for the TTSP from Section 2.7 for the MIP formulation and for the LP formulation of the QSAP. All constraints that we analyze can be written as linear inequalities of the form

$$\sum_{i \in M} \sum_{j \in N_i} \alpha_{ij} x_{ij} + \sum_{i \in M} \sum_{\substack{k \in M \\ i < k}} \sum_{j \in N_i} \sum_{l \in N_k} \beta_{ijkl} y_{ijkl} \leq d$$

or as a set of these linear inequalities.

## Lineshift Constraints

A lineshift constraint reduces the allowed shifts of a line. Thus, it corresponds to a reduction of the set of locations $L_i$ for an object $o_i$ and we get a smaller index set $N_i' \subseteq N_i$. This constraint can be introduced into the model by the coefficients

$$\alpha_{kl} = \begin{cases} 1 & \text{, if } k = i \text{ and } l \notin N_i' \\ 0 & \text{, else,} \end{cases}$$

$$\beta_{ijkl} = 0,$$

$$d = 0.$$

This leads to

$$\sum_{j \notin N_i'} x_{ij} \leq 0.$$

Combined with the nonnegativity constraints $x_{ij} \geq 0$ for all $j \in N_i$, we have $x_{ij} = 0$ for all $j \notin N_i'$

## Period Constraints

A period constraint enforces all solutions of a TTSP to keep a defined common meeting point. From the results of Section 2.7.4 we know that a period constraint can be formulated as a set of binary relations within the lines that belong to the common meeting point. Therefore, a period constraint defines a set of allowed quadruples $Q$, which represent the allowed line shift pairs.

To include a period constraint in the formulations of this chapter, we enforce all $y$-variables that belong to not allowed shift pairs to have a value of zero. This can be achieved with the coefficients

$$\alpha_{ij} = 0,$$

$$\beta_{ijkl} = \begin{cases} 1 & \text{, if } (i,j,k,l) \notin Q \\ 0 & \text{, else,} \end{cases}$$

$$d = 0,$$

which leads to the constraint

$$\sum_{i \in M} \sum_{\substack{k \in M \\ i < k}} \sum_{j \in N_i} \sum_{\substack{l \in N_k \\ (i,j,k,l) \notin Q}} y_{ijkl} \leq 0.$$

Combined with the nonnegativity constraint $y_{ijkl} \geq 0$ for all $i \in M$, $k \in M$ $(k > i)$, $j \in N_i$, $l \in N_k$, we get $y_{ijkl} = 0$ if $(i,j,k,l) \notin Q$.

**Connection Constraints**

The connection constraint is harder to formulate than the other two constraint types. Here, we have to differ between two cases. The first case occurs, e.g., if a convenience transfer must be preserved and there is only one possible departing tour. In this case, the allowed shift differences $s_i - s_k$ of the corresponding lines $l_i$ and $l_k$ define an interval $d_{\min} \leq s_i - s_k \leq d_{\max}$. Such an interval is visualized in Figure 3.1.

Figure 3.1:  Visualization of the allowed shift differences of a connection constraint. The rectangles mark forbidden shift differences.

This first version of the connection constraint can be formulated with two linear inequalities. The first one has the coefficients

$$\alpha_{pq} = \begin{cases} q & \text{, if } p = i \\ -q & \text{, if } p = k \\ 0 & \text{, else,} \end{cases}$$

$$\beta_{ijkl} = 0,$$

$$d = d_{\max}$$

and the second one has the coefficients

$$\alpha_{pq} = \begin{cases} -q & \text{, if } p = i \\ q & \text{, if } p = k \\ 0 & \text{, else,} \end{cases}$$

$$\beta_{ijkl} = 0,$$

$$d = -d_{\min}.$$

For a simplified notation, we examine an example where the allowed shift sets for both lines are identical ($N_i = N_k$) and without gaps. This leads to the constraint

$$\sum_{j \in N_i} j \cdot x_{ij} - \sum_{l \in N_k} l \cdot x_{kl} \leq d_{\max} \quad \wedge$$

$$\sum_{j \in N_i} j \cdot x_{ij} - \sum_{l \in N_k} l \cdot x_{kl} \geq d_{\min}.$$

If the lines have different sets $N_i$ and $N_k$, the coefficients $\alpha_{pq}$ must be changed accordingly so that the shift differences are calculated correctly. These changes depend on the waiting time in the current timetable and the borders of the corresponding transfer type.

The second case of a connection constraint occurs if, e.g., a forbidden almost transfer is considered or if the departing line has several possible tours that can be used for the regarded transfer. Here, the allowed shift differences cannot be formulated as a single interval but only as a set of intervals of the form

$$d_1 \leq s_i - s_k \leq d_2 \ \vee \ d_3 \leq s_i - s_k \leq d_4 \ \vee \ \dots$$

A visualization of the allowed shift differences in such a case is presented in Figure 3.2.

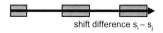

Figure 3.2:  Visualization of the allowed shift differences of a connection constraint. The rectangles mark forbidden shift differences. In this example, the allowed area consists of two intervals.

This type of constraint can be formulated with the help of disjunctions of conjunctions. These conjunctions are constructed as it was done in the first case. This leads to the constraint

$$
\Big( \sum_{j \in N_i} j \cdot x_{ij} - \sum_{l \in N_k} l \cdot x_{kl} \geq d_1 \ \wedge \sum_{j \in N_i} j \cdot x_{ij} - \sum_{l \in N_k} l \cdot x_{kl} \leq d_2 \Big) \vee
$$
$$
\Big( \sum_{j \in N_i} j \cdot x_{ij} - \sum_{l \in N_k} l \cdot x_{kl} \geq d_3 \ \wedge \sum_{j \in N_i} j \cdot x_{ij} - \sum_{l \in N_k} l \cdot x_{kl} \leq d_4 \Big) \vee
$$
$$
\ldots
$$

Here, each conjunction represents one allowed shift difference interval. An alternative approach is to directly place the constraint on the forbidden $y$-variables. In this case, if the shift combination pair $s_{ij}$ and $s_{kl}$ is forbidden, we introduce a constraint of the form

$$
y_{ijkl} \leq 0.
$$

## 3.6   QSAPs on graphs

Finding an optimal solution for a QSAP is very hard, especially for bigger problem sizes. Thus, in most cases one has to implement heuristics to find at least good solutions (that are not necessarily optimal). For this task, it is desirable to have a formulation of QSAPs on graphs, since the theory of heuristics for graph problems is widely studied.

This section deals with the aspects of formulating the QSAP on graphs. Furthermore, related graph problems and solution strategies for the QSAP are discussed.

### 3.6.1   Graphs - Definition and Notation

We start by introducing some basic definitions for graphs that we use in this thesis. Furthermore, we introduce the concept of multipartite graphs that is needed to define the QSAP on graphs.

**Definition 3.4 (Graph)**  *A graph is an ordered pair $G = (V, E)$, where $V$ defines a set of vertices and $E$ defines a set of edges, which are two-element subsets of $V$. For the QSAP, we name the vertices by two indices and the edges by four indices. Thus we have between the vertices $v_{ij}, v_{kl} \in V$ the corresponding undirected edge $e_{ijkl} = \{v_{ij}, v_{kl}\} \in E$.*

All graphs in this work are considered as undirected, meaning that the edges of these graphs have no orientation. For transferring the structure of a QSAP on graphs, we have to partition the vertices of a graph. A partition $\mathcal{V}$ of a set $V$ is a division of $V$ in non-overlapping and non-empty subsets of $V$ that cover $V$. Thus, we have $\mathcal{V} = \{V_1, ..., V_m\}$ with $V_i \subseteq V$ for all $i \in \{1, ..., m\}$, $\bigcup_{i \in M} V_i = V$ and $V_i \cap V_j = \emptyset$ if $i \neq j$.

If a partition has a special structure, meaning that the edges of the graph only occur between different vertex sets $V_i$ and $V_j$, we speak of a multipartite graph.

**Definition 3.5 (Multipartite graph)**  *A graph $G = (V, E)$ with a partition $\mathcal{V} = \{V_1, ..., V_m\}$ is called multipartite, if it satisfies*

$$\{v_{ij}, v_{il}\} \notin E \quad \forall i \in M, \; \forall v_{ij}, v_{il} \in V_i.$$

*Such a graph is called a complete multipartite graph, if in addition*

$$\{v_{ij}, v_{kl}\} \in E \quad \forall i, k \in M \; (i \neq k), \; \forall v_{ij} \in V_i, \; \forall v_{kl} \in V_k$$

*holds.*

Regarding a QSAP, these sets $V_i$ represent the sets $N_i$ from the QSAP and the single vertices $v_{ij}$ correspond to the variables $x_{ij}$.

Furthermore, we have to introduce the concept of cliques to represent the solutions of a QSAP on a graph. The formal definition is the following:

**Definition 3.6 (Clique, maximum Clique)**  *For a graph $G = (V, E)$, a vertex subset $C \subseteq V$ is called a clique, if*

$$\{v_1, v_2\} \in E \quad \forall v_1, v_2 \in C \quad (v_1 \neq v_2)$$

*holds. Such a clique is called a maximum clique of the graph $G$, if there exists no clique with more vertices in $G$.*

## 3.6.2   QSAP-Formulation on Graphs

After having defined the necessary requirements, we will introduce the concept of QSAPs on graphs in the following definition.

**Definition 3.7 (QSAP-graph)** *For a set $M = \{1, ..., m\}$ and the sets $N_i = \{1, ..., n_i\}$ with $i \in M$, the QSAP-graph*

$$G_{QSAP} = (V_{QSAP}, E_{QSAP})$$

*is the complete multipartite graph with the vertices*

$$V_{QSAP} = \{v_{ij} \mid i \in M, \ j \in N_i\}$$

*and the partition*

$$\mathcal{V} = \{V_1, ..., V_m\}, \ V_i = \{v_{i1}, ..., v_{in_i}\}.$$

*The edges of $G_{QSAP}$ are defined by*

$$E_{QSAP} = \{e_{ijkl} = \{v_{ij}, v_{kl}\} \mid v_{ij}, v_{kl} \in V_{QSAP} \ (i \neq k), \ j \in N_i, \ l \in N_k\}.$$

*The vertices $v_{ij}$ and the edges $e_{ijkl}$ have weights according to the object-location-cost-function (for the vertices) and the object-location-pair-cost-function (for the edges). These weights are given by*

$$w(v_{ij}) = f_{o_i}(l^i_j) \ and$$

$$w(e_{ijpq}) = f_{o_i,o_p}(l^i_j, l^p_q)$$

*with $f_{o_i}$ and $f_{o_i,o_p}$ from Definition 3.1.*

A small example of a QSAP-graph that corresponds to an assignment problem with four objects and three locations is presented in Figure 3.3.

For every feasible solution of an MIP or an LP formulation of a QSAP, we can construct a graph that corresponds to this solution. Such a solution graph $G_{sol}$ is a subgraph of $G_{QSAP}$.

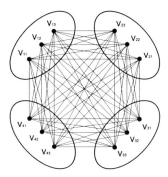

Figure 3.3: Graph $G_{QSAP}$ with $m = 4$ and $n_i = 3$ for all $i \in \{1, ..., 4\}$.

**Definition 3.8 (Solution graph)** *Given are a QSAP with the corresponding QSAP-graph $G_{QSAP} = (V_{QSAP}, E_{QSAP})$ and a solution of the MIP or LP formulation of this QSAP. The solution graph $G_{sol} = (V_{sol}, E_{sol})$ that corresponds to this solution is a subgraph of $G_{QSAP}$ with*

$$V_{sol} = \{v_{ij} \in V_{QSAP} \mid x_{ij} > 0\} \ and$$

$$E_{sol} = \{e_{ijkl} \in E_{QSAP} \mid y_{ijkl} > 0\}.$$

*The variables $x_{ij}$ and $y_{ijkl}$ are the assignment decisions of the MIP or the LP solution.*

The following proposition helps to simplify the problem formulation by removing the unnecessary vertex weights.

**Proposition 3.9** *In general, the vertex weights of a QSAP-graph $G_{QSAP}$ can be eliminated by transferring them to the adjacent edges. A possibility to do this is to redistribute the vertex weights of vertex $v_{ij}$ to the edges between $v_{ij}$ and the vertices of the vertex set $V_k$ ($i \neq k$) as follows:*

$$w(e_{ijk1}) \leftarrow w(e_{ijk1}) + w(v_{ij})$$

$$\ldots$$

$$w(e_{ijkn_k}) \leftarrow w(e_{ijkn_k}) + w(v_{ij})$$

$$w(v_{ij}) \leftarrow 0.$$

*This redistribution does not change the solution value of any feasible solution of a QSAP.*

**Proof:**

For a vertex $v_{ij}$ and the corresponding variable $x_{ij}$, we know from the constraint

$$x_{ij} = \sum_{l=1}^{n_k} y_{ijkl} \quad (k \in M \setminus \{i\})$$

of the MIP and the LP formulation that we can redistribute the objective function value to the edges. This is the case, because of

$$b_{ij} x_{ij} = \sum_{l=1}^{n_k} b_{ij} y_{ijkl}.$$

Thus, transferring the coefficient $b_{ij}$ to the coefficients $c_{ijkl}$ $(l = 1, ..., n_k)$ does not change the structure of the objective function.

$\square$

As a consequence of this proposition, we only consider edge weights in most applications in this thesis. Therefore, we treat the vertex weights as being all equal to zero.

In the following lemma, we introduce an important relation between the theory of QSAPs and graph theory.

**Lemma 3.10** *The problem of finding the minimal solution of a QSAP is equivalent to finding an m-clique with a minimal sum of vertex weights and edge weights in the corresponding QSAP-graph.*

**Proof:**

We can easily construct a bijection $g$ between the set of the solutions of a QSAP and the $m$-cliques of the corresponding QSAP-graph. Here, a solution $x$ of the QSAP with $x_{ij} \in \{0, 1\}$ for all $i \in M$ and for all $j \in N_i$ is mapped by this bijection to an $m$-clique $C$ in $G_{QSAP}$. We have $v_{ij} \in g(x) = C$ if and only if $x_{ij} = 1$. This mapping is injective, since two different solutions $x_1$ and $x_2$ are mapped to different $m$-cliques $C_1$ and $C_2$. It is in addition surjective, since there exists a unique solution $x$ of the QSAP for each $m$-clique $C$ so that $g(x) = C$ holds.

The objective function coefficients $b_{ij}$ are integrated in the graph model as vertex weights $w(v_{ij})$ and the coefficients $c_{ijkl}$ are integrated as edge weights. Thus, the product $b_{ij} \cdot x_{ij}$ is counted in the objective function if and only if the vertex $v_{ij}$ with weight $w(v_{ij}) = b_{ij}$ is part of the corresponding $m$-clique.

The same holds for the product $c_{ijkl} \cdot x_{ij} \cdot x_{kl}$ that is counted if and only if the edge $e_{ijkl} = \{v_{ij}, v_{kl}\}$ (that has the weight $w(e_{ijkl}) = c_{ijkl}$) is part of the $m$-clique. Hence, the objective function value of solution $x$ is the same as the sum of the vertex weights and the edge weights of the corresponding $m$-clique. Thus, the minimal solution of the QSAP has the same solution value as the $m$-clique whose sum of the vertex weights and edge weights is minimal.

$\square$

The objective function of the QSAP can be converted to a form

$$f(x) = \sum_{i=1}^{n} \sum_{j=1}^{n_i} 0 \cdot x_{ij} + \sum_{i=1}^{m-1} \sum_{k=i+1}^{m} \sum_{j=1}^{n_i} \sum_{l=1}^{n_k} \overline{c}_{ijkl} x_{ij} x_{kl}$$

so that only coefficients $\overline{c}_{ijkl}$ are needed and all coefficients $b_{ij}$ have value zero. This was shown in Proposition 3.9. Thus, we know that edge weights are sufficient for a complete characterization of the objective function on graphs.

For the QSAP-graph $G_{QSAP}$, the influence of constraints corresponds to a removal of vertices and edges. An example of a constrained QSAP-graph can be seen in Figure 3.4.

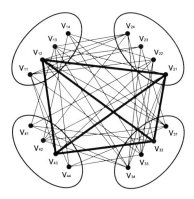

Figure 3.4: Constrained QSAP-graph with $m = 4$ and $n_i = 3 \; \forall i \in \{1, ..., 4\}$. The bold edges show a clique that represents a feasible solution.

## 3.7   QAP-Formulation on Graphs

According to the previous theory of QSAPs on graphs, we introduce in this section the related theory of QAPs on graphs. We denote the graph that corresponds to this type of problem by $G_{QAP}$. The following definition introduces the structure of this graph.

**Definition 3.11 (QAP-graph)** *For a set $M = \{1, ..., m\}$, the QAP-graph $G_{QAP} = (V_{QAP}, E_{QAP})$ is the multipartite graph with the vertex set*

$$V_{QAP} = \{v_{ij} \mid i, j \in M\}$$

*and the partition*

$$\mathcal{V} = \{V_1, ..., V_m\},\ V_i = \{v_{i1}, ..., v_{im}\}.$$

*The edges of $G_{QAP}$ are defined by*

$$E_{QAP} = \{e_{ijkl} = \{v_{ij}, v_{kl}\} \mid v_{ij}, v_{kl} \in V_{QAP}\ (i < k),\ j, l \in M\ (j \neq l)\}.$$

*The vertices $v_{ij}$ and the edges $e_{ijkl}$ have weights corresponding to the object-location-cost-function (for the vertices) and the object-location-pair-cost-function (for the edges). These weights are given by*

$$w(v_{ij}) = f_{o_i}(l_j)\ and$$

$$w(e_{ijst}) = f_{o_i, o_s}(l_j^i, l_t^s).$$

Figure 3.5 shows an example of a small QAP-graph that corresponds to a QAP with four objects and four locations.

In the following proposition, we show that the constrained QSAP is a generalization of the QAP.

**Proposition 3.12** *The QAP is a special case of the constrained QSAP.*

**Proof:**

Given is a QAP with $m$ objects. We construct a constrained QSAP of size $M = \{1, ..., m\}$, $N_i = \{1, ..., m\}$ for all $i \in M$ and the corresponding graph $\hat{G} = (\hat{V}, \hat{E})$ with the vertex set

$$\hat{V} = V_{QSAP} = \{v_{ij} \mid i \in M, j \in N_i\}$$

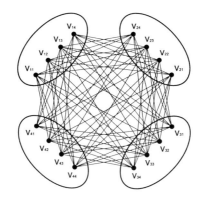

Figure 3.5: QAP-Graph $G_{QAP}$ with $m = 4$.

and the edge set $\hat{E}$, where all edges $e_{ijkl}$ with $j = l$ are removed by constraints. Thus, we have

$$\hat{E} = \{e_{ijkl} \in E_{QSAP} \mid i < k, j \neq l\}.$$

The graph $\hat{G}$ is equivalent to the QAP-graph $G_{QAP}$ from Definition 3.11. Thus, each $m$-clique of the QAP-graph can be mapped one-to-one to an $m$-clique in $\hat{G}$. By solving the QAP, we get a solution of the constrained QSAP and vice versa.

□

The QAP is widely studied and a lot of research on this topic was made in the last years. It is "arguably the hardest of the NP-hard problems" (Wolkowicz, [63]), since it is very hard to solve small problem sizes exactly or to calculate good lower bounds. The QAP is a special case of the constrained QSAP. Thus, it is plausible to use those approaches for the QSAP that generate promising results for the QAP.

## 3.8 Semi-Definite Programming vs. Reformulation Linearization Technique

The QSAP is NP-hard, so the optimal solution is not computable in a reasonable time for bigger problem instances. Thus, we have to look for different approaches. Metaheuristics are a good choice to find good solutions, but there is always the uncertainty how far one is still away from the optimal solution. Here, generating good lower bounds can reduce this uncertainty, since the bounds allow an estimation of how good the generated solutions of the metaheuristics are.

There are currently two state-of-the-art methods for generating lower bounds for the QAP. The first one is a Semi-Definite Programming (SDP) approach and the second one is the Reformulation Linearization Technique (RLT). Both techniques can also be applied to the QSAP and we discuss their advantages and disadvantages in this section.

Table 3.8 presents first results of some SDP and RLT approaches that provide best lower bounds for several test instances from QAPLIB [10]. QAPLIB is a QAP Library that contains several test instances of different sizes. The columns of the table correspond to

- Opt: optimal solution of the problem,

- RS03: SDP bound from Rendl and Sotirov [46],

- R04: SDP bound from Roupin [47],

- BV04: lift-and-project SDP bound from Burer and Vandenbussche [9],

- HH01: Hahn-Hightower RLT-2 dual ascent bound from Adams et al. [1] (remark that the bound was first published in 2001, but the presented bounds were recently calculated in 2007),

- HZ07: Hahn-Zhu RLT-3 bound from Zhu [65].

Tight lower bounds can be found by the SDP approach and the RLT approach for several of these instances. The question which of these approaches is the best one cannot be answered and is clearly problem dependent. Furthermore, there is not only the problem of just finding good bounds. It is also important that the computation of these bounds is efficient.

## 3.8.1   Semi-Definite Programming

SDP deals with optimizing a linear objective function over symmetric positive semidefinite matrix variables under linear constraints. SDP is a subfield of convex optimization and the formulation can be solved efficiently by interior

| Instance | Opt | RS03 | R04 | BV04 | HH01 | HZ07 |
|---|---|---|---|---|---|---|
| Had16 | 3720 | 3699 | 3720** | 3672 | 3720* | 3719.1** |
| Had18 | 5358 | 5317 | 5356 | 5299 | 5358* | 5357** |
| Had20 | 6922 | 6885 | 6920 | 6811 | 6922* | 6920 |
| Kra30a | 88900 | 77647 | | 86678 | 86247 | |
| Kra30b | 91420 | 81156 | | 87699 | 87107 | |
| Nug12 | 578 | 557 | 567 | 568 | 578* | 577.2** |
| Nug15 | 1150 | 1122 | 1138 | 1141 | 1150* | 1149.1** |
| Nug18 | 1930 | | | | | 1930* |
| Nug20 | 2570 | 2451 | 2494 | 2506 | 2508 | 2568.1** |
| Nug22 | 3596 | | | | 3511 | 3594.04** |
| Nug30 | 6124 | 5803 | | 5934 | 5770 | |
| Rou15 | 354210 | 333287 | 348838 | 350207 | 354210* | 354210* |
| Rou20 | 725520 | 663833 | 691124 | 695123 | 699390 | 725314.4 |
| Tai20a | 703482 | 637300 | | 671685 | 675870 | 703482* |
| Tai25a | 1167256 | 1041337 | | 1112862 | 1091653 | |
| Tai30a | 1818146 | 1652186 | | 1706875 | 1686290 | |
| Tho30 | 149936 | 136059 | | 142814 | 136708 | |

Table 3.1: Fragment of the best lower bound matrix for different QAP instances from QAPLib.
RS03, R04, BV04 are SDP methods, HH01, HZ07 are RLT methods
\* Problem solved exactly by lower bound calculation
\*\* Optimum verified by bound calculation

point methods. Two examples of an application of SDP for QAPs can be found in [12] and [64]. These SDP approaches for QAPs are based on a QAP formulation of the form

$$(QAP) \quad \min_{X \in \Pi} tr(AXBX^T - 2CX^T),$$

where $A$, $B$ and $C$ are $m \times m$ matrices, $tr$ defines the trace of a matrix (the sum of the elements of the main diagonal) and $\Pi$ is the set of all permutation matrices of size $m$.

This type of QAP formulation has the disadvantage that it requires a decomposition of the objective function. Facility location problems have, for example, such a decomposition. In this case, the cost function is constructed from a distance component for the locations and a transport volume that must be transported between the different facilities. Here, the variables $c_{ijkl}$ can be written as a product $c_{ijkl} = \overline{c}_{ik} \cdot \widehat{c}_{kl}$.

For the problem of this thesis, such a decomposition of the QSAP is not possible, since the objective function directly consists of elements that depend on lines and shifts at the same time. Thus, state-of-the-art SDP techniques cannot be applied to our type of QSAP, because of these non-decomposable $c_{ijkl}$ coefficients in our objective function.

### 3.8.2   Reformulation Linearization Technique

The RLT approach is a new technique for generating tight linear or convex relaxations for both discrete and continuous nonconvex programming problems. It consists generally of two steps. A reformulation step generates new non-linear valid inequalities. Afterwards, in the linearization step, the product terms in these inequalities are replaced by new continuous variables. A good overview for RLT approaches for global optimization can be found in [57].

With this technique, the approach generates stepwise tighter polyhedral formulations of the original relaxed problem. The projection of the higher dimensional polyhedra to the space of the original $x$-variables and $y$-variables of the LP formulation of the QAP gives a stepwise improving polyhedral formulation

$$X_{P_0} \supseteq X_{P_1} \supseteq X_{P_2} \supseteq ... \supseteq X_{P_n} = \text{conv}(X).$$

Here, $X_{P_0}$ is the polyhedron that corresponds to the relaxed version of the original IP formulation and $conv(X)$ is the convex hull of the feasible solutions of the original QAP.

In Chapter 5, we discuss the theory of the RLT and its application to the QSAP in detail. In addition, we try to determine the smallest RLT level $t \leq n$ so that $X_{P_t} = \text{conv}(X)$ holds.

## 3.9   Polyhedral Theory

This section deals with the theory of describing QSAPs as polytopes. We extend the work of Saito et al. [49] and we present some new results for our generalized version of the QSAP.

## 3.9.1 The QSAP Polytope

We begin this section by introducing the definition of polyhedrons and polytopes from Nemhauser and Wolsey [42].

**Definition 3.13 (Polyhedron, Polytope, Extreme Point)** *A polyhedron $P \subseteq \mathbb{R}^n$ is a set of points that satisfy a finite number of linear inequalities; that is, $P = \{x \in \mathbb{R}^n : Ax \leq b\}$, where $(A, b)$ is an $m \times (n + 1)$ matrix. Thus, a polyhedron is an intersection of half-spaces. A polytope is a bounded polyhedron.*

*A point $x \in P$ is an extreme point of $P$, if there exist no two points $x_1, x_2 \in P$, $x_1 \neq x_2$, so that $x = \frac{1}{2}x_1 + \frac{1}{2}x_2$ holds.*

In addition, we need the concept of convex combinations of points in a polyhedron.

**Definition 3.14 (Convex Combination, Convex Hull)** *For a set $S \subseteq \mathbb{R}^n$, a point $x \in \mathbb{R}^n$ is a convex combination of points of $S$, if there exists a finite set of points $\{x_i\}_{i=1}^t$ in $S$ and a $\lambda \in \mathbb{R}_+^t$ with $\sum_{i=1}^t \lambda_i = 1$ and*

$$x = \sum_{i=1}^t \lambda_i x_i.$$

*The convex hull of $S$, denoted by $conv(S)$, is the set of all points that are convex combinations of points in $S$.*

The polytope of the QSAP of size $M = |m|$ and $N = \{N_1, ..., N_m\}$ is denoted by $QSAP_{m,N}$. This notation is chosen according to the $QSAP_{m,n}$ polytope of Saito et al. from [49]. Here, the authors analyze the QSAP that arises if all sets $N_i$ are of the same size ($|N_i| = n$ for all $i \in M$).

To define our new polytopes, we need the theory of describing QSAPs on graphs (cf. Subsection 3.6). For a QSAP-graph $G_{QSAP} = (V, E)$, we introduce the following notation:

For a subset $V' \subseteq V$, we define

$$E(V') := \{\{u, v\} \in E \mid u, v \in V'\}\}$$

and for the vertex set $V = \{v_1, v_2, ...\}$, the characteristic vector $\chi_{V'} \in \{0, 1\}^{|V|}$ is defined by

$$\chi_{V'}^i := \begin{cases} 1 & \text{if } v_i \in V', \\ 0 & \text{else,} \end{cases} \quad (i \in \{1, ..., |V|\}).$$

Accordingly, for the edge set $E = \{e_1, e_2, ...\}$ and a subset $E' \subseteq E$, the characteristic vector $\chi_{E'} \in \{0,1\}^{|E|}$ is defined by

$$\chi_{E'}^i := \begin{cases} 1 & \text{if } e_i \in E', \\ 0 & \text{else,} \end{cases} \quad (i \in \{1, ..., |E|\}).$$

These characteristic vectors are used to define the QSAP polytope.

**Definition 3.15 (QSAP polytope)** *The convex hull*

$$QSAP_{m,N} = conv\{(\chi_C, \chi_{E(C)}) \in \mathbb{R}^{|V|} \times \mathbb{R}^{|E|} \mid C \text{ is an } m\text{-clique in } G\}.$$

*defines the QSAP polytope* $QSAP_{m,N}$.

To simplify the further work with QSAP-graphs, we introduce the following notation. For a vertex set $V' \subseteq V$, we have

$$x(V') := \sum_{v \in V'} x_v \quad \text{and} \quad y(V') := \sum_{e \in E(V')} y_e.$$

For an edge set $E' \subseteq E$, we define in addition

$$y(E') := \sum_{e \in E'} y_e.$$

Furthermore, for disjoint sets $S, T \subseteq V$, we introduce the shorter notation

$$E(S : T) := \{\{u, v\} \in E \mid u \in S, v \in T\}$$

for the edges between the sets $S$ and $T$. Henceforth, we use the shorter notation

$$y(S : T) := y(E(S : T)).$$

Finally, we define for every $i \in M$ the set

$$\text{row}_i := \{v_{ij} \in V_{QSAP} \mid j \in N_i\}.$$

This last definition leads to a rectangular form of the QSAP-graphs that we present in Figure 3.6.

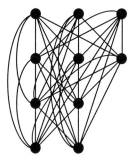

Figure 3.6: QSAP-graph with $m = 4$, $|N_1| = |N_2| = 3$ and $|N_3| = |N_4| = 2$ visualized in rectangular form.

With this notation, we can reformulate the already defined MIP from Section 3.4 with a QSAP-graph $G_{QSAP} = (V, E)$ as follows:

$$(MIP_2) \ \min \sum_{v \in V} a_v \cdot x_v + \sum_{e \in E} b_e \cdot y_e$$

$$s.t. \ x(row_i) = 1 \quad \forall i \in M, \tag{3.1}$$

$$-x_v + y(v : row_k) = 0 \quad \forall v \in V, \ \forall k \in M \ (v \notin row_k), \tag{3.2}$$

$$x_v \in \{0, 1\} \quad \forall v \in V, \tag{3.3}$$

$$y_e \geq 0 \quad \forall e \in E. \tag{3.4}$$

Note that the constraint (3.2) is equivalent to

$$-x_{ij} + y(v_{ij} : row_k) = 0 \quad \forall v_{ij} \in V, \ \forall k \in M \ (i \neq k).$$

A connection between the $(MIP_2)$ formulation and the extreme points of the $QSAP_{m,N}$ polytope is presented in the following proposition.

**Proposition 3.16** *A vector $(x, y) \in \mathbb{R}^{|V|} \times \mathbb{R}^{|E|}$ is an extreme point of $QSAP_{m,N}$ if and only if it satisfies (3.1), (3.2), (3.3) and (3.4).*

**Proof:**
On the one hand, let $(x, y) \in \mathbb{R}^{|V|} \times \mathbb{R}^{|E|}$ be an extreme point of $QSAP_{m,N}$. Then clearly $(3.1) - (3.4)$ hold.

On the other hand, let $(3.1) - (3.4)$ be true for a vector $(x, y) \in \mathbb{R}^{|V|} \times \mathbb{R}^{|E|}$. The constraints $(3.1) - (3.4)$ imply that $y_e \in \{0, 1\}$ holds (cf. Proposition

3.2). In addition, the constraints (3.1) enforce that exactly $m$ vertices have a value $x_v = 1$. The equations

$$-x_{ij} + y(v_{ij} : row_k) = 0 \quad \forall j \in N_i$$
$$-x_{kl} + y(v_{kl} : row_i) = 0 \quad \forall l \in N_k$$

ensure that $y_{ijkl} = x_{ij} \cdot x_{kl}$ holds for every edge $e_{ijkl} = \{v_{ij}, v_{kl}\}$ with $(i \neq k)$. Hence, only edges between nodes $v_{ij}$ and $v_{kl}$ with $x_{ij} = x_{kl} = 1$ have a value $y_e > 0$. We get that

$$C := \{v_{ij} \in V \mid x_{ij} = 1\}$$

is an $m$-clique in $G_{QSAP}$ and $E(C) = \{e \in E \mid y_e = 1\}$ are the edges inside this clique. Thus, $(x, y) \in \mathbb{R}^{|V|} \times \mathbb{R}^{|E|}$ is an extreme point of $QSAP_{m,N}$, since this point cannot be a linear combination of two other points in $G_{QSAP}$.

$\square$

## 3.9.2   The Star-Transformation

The star-transformation that we introduce in this section is an isomorphic projection of the $QSAP_{m,N}$ polytope to obtain a full-dimensional polytope that is essentially equivalent to the original one. The transformation is a modified version of the star-transformation from Saito et al. [49]. The first research on the star-transformation was done by Jünger and Kaibel in [29]. Here, the authors introduce this transformation to analyze the properties of the polytope for the QAP.

The idea of this transformation is that no information is lost if one vertex is removed from each set $row_i$. In our case, we always remove the first vertex $v_{i1}$. Thus, if we regard the $(n_i - 1)$ vertices of row $i$ that remain after such a removal, there are two possibilities. If one of the vertices $v_{ij}$ is chosen ($\exists j \in \{2, ..., n_i\}$ : $x_{v_{ij}} = 1$), then the variable that corresponds to the removed vertex $x_{v_{i1}}$ must have value zero. But if no vertex is chosen ($\forall j \in \{2, ..., n_i\}$ : $x_{v_{ij}} = 0$), then we know that the removed vertex $v_{i1}$ must have the corresponding variable assignment $x_{v_{i1}} = 1$. For the notation of such a reduction, we introduce the reduced index sets

$$N_i^* = N_i \setminus \{1\} = \{2, ..., n_i\}$$

for all $i \in M$. With these new sets, it is still possible to represent all extreme points of the polytope $QSAP_{m,N}$. The graph $G_{QSAP} = (V, E)$ is transformed into the smaller graph $G^* = (V^*, E^*)$ with

$$V^* := \{v_{ij} \in V \mid i \in M, j \in N_i^*\} \text{ and}$$
$$E^* := \{\{v_{ij}, v_{kl}\} \in E \mid i, k \in M, i \neq k, j \in N_i^*, l \in N_k^*\}.$$

An example of a star-transformation is presented in Figure 3.7. Here, a QSAP-graph is shown to which the star transformation is applied.

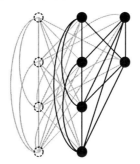

Figure 3.7: Graph $G^* = (V^*, E^*)$ (black) with $|L| = 4$, $|N_1^*| = |N_2^*| = 2$ and $|N_3^*| = |N_4^*| = 1$ that is constructed by the star transformation. The removed vertices of the original QSAP-graph are marked as dashed circles and the removed edges are shown as gray lines.

In such a reduced graph $G^*$, we now search for all types of cliques instead of focusing only on $m$-cliques. If we regard, for example, the solution, in which $v_{i1}$ is chosen for all $i \in M$ ($x_{v_{i1}} = 1$ for all $i \in M$), then the empty clique (a clique with zero vertices) corresponds to this solution in $G^*$. Therefore, we construct a new polytope as the convex hull over all possible cliques in $G^*$ instead of using only $m$-cliques.

**Definition 3.17** *The $QSAP^*_{m,N^*}$ polytope is defined by*

$$QSAP^*_{m,N^*} = conv\{(\chi_C, \chi_{E^*(C)}) \in \mathbb{R}^{|V^*|} \times \mathbb{R}^{|E^*|} \mid C \text{ is a clique in } G^*\}$$

*with adjusted characteristic vectors $\chi$ ($\chi_C \in \{0,1\}^{|V^*|}$, $\chi_{E^*(C)} \in \{0,1\}^{|E^*|}$). For this formulation, we specify that an empty set is a clique of size zero.*

In comparison to the $QSAP_{m,N}$ polytope, the $QSAP^*_{m,N^*}$ polytope has the advantage that it is full-dimensional. This is shown in the following proposition.

**Proposition 3.18** *The $QSAP^*_{m,N^*}$ polytope is full-dimensional, i.e.,*

$$\dim(QSAP^*_{m,N^*}) = |V^*| + |E^*|$$

*holds.*

**Proof:**

On the one hand, the inequality

$$\dim(QSAP^*_{m,N^*}) \leq |V^*| + |E^*|$$

comes from the definition of the polytope. On the other hand, we have $(|V^*| + |E^*| + 1)$ vectors

$$(0,0), \quad (1 \text{ vector})$$
$$(\chi_u, 0) \, \forall u \in V^*, \quad (|V^*| \text{ vectors})$$
$$(\chi_u + \chi_v, \chi_{uv}) \, \forall \{u,v\} \in E^* \quad (|E^*| \text{ vectors})$$

that are affinely independent in $QSAP^*_{m,N^*}$. Thus,

$$|V^*| + |E^*| \leq \dim(QSAP^*_{m,N^*})$$

holds. This proves the proposition.

$\square$

With this knowledge, we want to examine the dimension of the initial $QSAP_{m,N}$ polytope. Therefore, we need the following affine subspaces:

$$\mathcal{A} := \left\{ (x,y) \in \mathbb{R}^{|V|} \times \mathbb{R}^{|E|} \,\middle|\, \begin{array}{c} x(row_i) = 1 \quad \forall i \in M, \\ -x_{ij} + y(v_{ij} : row_k) = 0 \\ \forall i,k \in M \ (i \neq k), \ \forall j \in N_i \end{array} \right\},$$
$$\mathcal{U} := \left\{ (x,y) \in \mathbb{R}^{|V|} \times \mathbb{R}^{|E|} \,\middle|\, x_v = 0 \quad \forall v \in V \setminus V^*, \ y_e = 0 \quad \forall e \in E \setminus E^* \right\}.$$

For the further results, we introduce the following notation:

$$row_i^* := row_i \setminus \{v_{i1}\}, \ row(v_{ij}) := row_i, \ row^*(v_{ij}) := row_i^*.$$

Saito et al. constructed an affine mapping

$$\phi : \mathbb{R}^{|V|} \times \mathbb{R}^{|E|} \to \mathbb{R}^{|V|} \times \mathbb{R}^{|E|}$$

between these two sets $\mathcal{A}$ and $\mathcal{U}$ and showed that $\phi : \mathcal{A} \to \mathcal{U}$ is an affine isomorphism. This affine mapping $\phi$ is defined by

$$\phi(x,y)_v := \begin{cases} x_v & \text{, if } v \in V^*, \\ x_v - (1 - x(row^*(v))) & \text{, if } v \in V \setminus V^* \end{cases}$$

$$\phi(x,y)_e := \begin{cases} y_e & \text{, if } e = \{u,v\} \in E^*, \\ y_e - (x_u - y(u : row^*(v))) & \text{, if } e \in E(V^* : V \setminus V^*), \\ y_e - (1 - x(row^*(u) \cup row^*(v)) \\ \quad + y(row^*(u) \cup row^*(v))) & \text{, if } e \in E(V \setminus V^*). \end{cases}$$

It is easy to see that $\phi$ is an affine isomorphism. By multiplying the matrix that corresponds to the affine mapping with a permutation matrix, we can construct a lower triangular matrix with only diagonal elements with value one. This permutation matrix changes the formulation so that the removed vertices $v_{i1}$ (for all $i \in M$) are treated as being the last vertex of $row_i$. The diagonal structure directly implies that $\phi$ is an affine isomorphism.

For the QSAP polytopes, we have

$$QSAP_{m,N} \subseteq \mathcal{A} \text{ and}$$

$$QSAP^*_{m,N^*} \subseteq \mathbb{R}^{|V^*|} \times \mathbb{R}^{|E^*|} \cong \mathcal{U}.$$

In addition, we define the linear mapping

$$\iota : \mathbb{R}^{|V^*|} \times \mathbb{R}^{|E^*|} \to \mathbb{R}^{|V|} \times \mathbb{R}^{|E|}$$

by

$$(\iota(x,y))_v := \begin{cases} x_v & \text{if } v \in V^*, \\ 0 & \text{if } v \in V \setminus V^*, \end{cases}$$

and

$$(\iota(x,y))_e := \begin{cases} y_e & \text{if } e \in E^*, \\ 0 & \text{if } e \in E \setminus E^*. \end{cases}$$

With this linear mapping, we directly get

$$\iota(QSAP^*_{m,N*}) \subseteq \phi(QSAP_{m,N}).$$

Thus,

$$\dim(QSAP^*_{m,N*}) \leq \dim(\phi(QSAP_{m,N})) \qquad (3.5)$$

holds. With this inequality, we can prove the following new theorem.

**Theorem 3.19** *For the dimension of the $QSAP_{m,N}$-polytope, the equation*

$$\dim(QSAP_{m,N}) = \dim(\mathbb{R}^{|V|} \times \mathbb{R}^{|E|}) - \left( \frac{3m - m^2}{2} + (m-1) \cdot \sum_{i \in M} n_i \right)$$

*holds.*

**Proof:**

Since $\phi$ is an affine isomorphism between $\mathcal{A}$ and $\mathcal{U}$, we get $\dim(\mathcal{A}) = \dim(\mathcal{U})$. For the dimension of $\mathcal{U}$, we have

$$\dim(\mathcal{U}) = \sum_{i \in M}(n_i - 1) + \sum_{\substack{i,k \in M \\ i < k}}(n_i - 1)(n_k - 1)$$

$$= \sum_{i \in M} n_i + \sum_{\substack{i,k \in M \\ i < k}}(n_i \cdot n_k) - \left( m - \sum_{\substack{i,k \in M \\ i < k}} 1 + \sum_{\substack{i,k \in M \\ i < k}}(n_i + n_k) \right)$$

$$= \dim(\mathbb{R}^{|V|} \times \mathbb{R}^{|E|}) - \left( m - \frac{m \cdot (m-1)}{2} + (m-1) \cdot \sum_{i \in M} n_i \right)$$

$$= \dim(\mathbb{R}^{|V|} \times \mathbb{R}^{|E|}) - \left( \frac{3m - m^2}{2} + (m-1) \cdot \sum_{i \in M} n_i \right)$$

With this equation, inequality (3.5), the affine isomorphism $\phi$ and Proposition 3.18, we get

$$\dim(\mathcal{A}) = \dim(\mathcal{U})$$

$$= \dim\left(\mathbb{R}^{|V|} \times \mathbb{R}^{|E|}\right) - \left( \frac{3m - m^2}{2} + (m-1) \cdot \sum_{i \in M} n_i \right)$$

$$= \dim\left(\mathbb{R}^{|V^*|} \times \mathbb{R}^{|E^*|}\right)$$

$$= \dim(QSAP^*_{m,N^*})$$

$$\leq \dim(\phi(QSAP_{m,N}))$$

$$= \dim(QSAP_{m,N})$$

$$\leq \dim(\mathcal{A}).$$

$\square$

Theorem 3.19 is a generalization of the result from Saito et al., who proved in [49] a formula for the dimension of the $QSAP_{m,N}$ polytope with $|N_i| = n$

for all $i \in M$. This polytope is denoted by $QSAP_{m,n}$. This special case is shown in the following proposition.

**Proposition 3.20** *If all sets $N_i$ are of equal size ($|N_i| = n$ for all $i \in M$), we get the result of Saito et al., namely*

$$\dim(QSAP_{m,n}) = \dim(\mathbb{R}^{|V|} \times \mathbb{R}^{|E|}) - \left( m + \frac{1}{2}m(m-1)(2n-1) \right)$$

*for the dimension of the $QSAP_{m,n}$ polytope.*

**Proof:**

Inserting $n_i = n$ in the result of Theorem (3.19) yields

$$\dim(QSAP_{m,n}) = \dim(\mathbb{R}^{|V|} \times \mathbb{R}^{|E|}) - \left( \frac{3m - m^2}{2} + (m-1) \cdot \sum_{i \in M} n_i \right)$$

$$= \dim(\mathbb{R}^{|V|} \times \mathbb{R}^{|E|}) - \left( \frac{3m - m^2}{2} + (m-1) \cdot m \cdot n \right)$$

$$= \dim(\mathbb{R}^{|V|} \times \mathbb{R}^{|E|}) - \left( m - \frac{m(m-1)}{2} + \frac{m(m-1)2n}{2} \right)$$

$$= \dim(\mathbb{R}^{|V|} \times \mathbb{R}^{|E|}) - \left( m + \frac{1}{2}m(m-1)(2n-1) \right).$$

$\square$

### 3.9.3 Defining Facets for the QSAP Polytope

With the results for the QSAP polytopes from the previous subsection, it stands to reason to search for facets and facet defining inequalities of these inequalities. In this subsection, we will give a short overview of the results of Saito et al. on facets. These results are also valid for our generalized form of the QSAP.

Before presenting the first results, we need the definition of valid inequalities.

**Definition 3.21 (Valid Inequality)** *Given an $m \times n$ matrix $A$ and a vector $b$, an inequality $Ax \leq b$ is called valid for a set of points $Y$, if for every $y \in Y$ the inequality $Ay \leq b$ holds.*

With this definition of valid inequalities, we define faces and facets for polyhedrons.

**Definition 3.22 (Face, Facet)** *A face $F$ of a polyhedron $P$ is a subset $F \subseteq P$ with $F = \{p \in P \mid Ap = b\}$ for a valid inequality $Ax \leq b$. We call such a face a facet, if it satisfies $\dim(F) = \dim(P) - 1$.*

*For a face $F$ of a polyhedron $P$ with $F = \{p \in P \mid Ap = b\}$, the valid inequality $Ax \leq b$ is said to define the face.*

The following proposition introduces first facets for the $QSAP^*_{m,N^*}$ polytope.

**Proposition 3.23** *For the $QSAP^*_{m,N^*}$ polytope, the inequalities*

$$y_e \geq 0 \quad \forall e \in E^*, \tag{3.6}$$

$$-x_v + y(v : row^*_k) \leq 0 \quad \forall v \in V^*, \ \forall k \in M \ (v \notin row_k), \tag{3.7}$$

$$x(row^*_i \cup row^*_k) - y(row^*_i \cup row^*_k) \leq 1 \quad \forall i, k \in M \ (i < k) \tag{3.8}$$

*define facets.*

**Proof:**

The validity of each inequality is easy to see. To prove that the inequalities are facets, we present $\dim(QSAP^*_{m,N^*}) = (|V^*| + |E^*|)$ affinely independent vectors from the corresponding face of each inequality.

(3.6):
$(0,0)$, (1 vector)
$(\chi_v, 0) \ \forall v \in V^*$, ($|V^*|$ vectors)
$(\chi_u + \chi_v, \chi_{uv}) \ \forall \{u, v\} \in E^* \setminus \{e\}$, ($|E^*| - 1$ vectors)

(3.7) ($v = v_{ij}$, $k$ fixed):
$(0,0)$, (1 vector)
$(\chi_u, 0) \ \forall u \in V^* \setminus \{v\}$, ($|V^*| - 1$ vectors)
$(\chi_v + \chi_w, \chi_{vw}) \ \forall w \in row^*_k$, ($|row^*_k|$ vectors)
$(\chi_{v_1} + \chi_{v_2}, \chi_{v_1 v_2}) \ \forall \{v_1, v_2\} \in E^*(V^* \setminus \{v\})$,
  ($|E^*| - \sum_{j \in M \setminus \{i\}} |row^*_j|$ vectors)
$(\chi_u + \chi_v + \chi_w, \chi_{uv} + \chi_{uw} + \chi_{vw}) \ \forall u \notin row^*_k \cup row^*_i, \exists w \in row^*_k$
  ($\sum_{t \in M \setminus \{i,k\}} |row^*_t|$ vectors)

(3.8) $(T = row_i^* \cup row_k^*)$:

$(\chi_v, 0)$ $\forall v \in T$, $(|row_i^*| + |row_k^*|$ vectors)

$(\chi_v + \chi_w, \chi_{vw})$ $\forall w \in V^* \setminus T, \exists v \in T,$

$\quad (\sum_{s \in M \setminus \{i,k\}} (|row_s^*|)$ vectors)

$(\chi_v + \chi_w, \chi_{vw})$ $\forall (v, w) \in E^*(T)$, $(|row_i^*| \cdot |row_k^*|$ vectors)

$(\chi_u + \chi_v + \chi_w, \chi_{uv} + \chi_{uw} + \chi_{vw})$ $\forall \{u, v\} \in E^*(T : V^* \setminus T), \exists v \in T,$

$\quad (\sum_{s \in M \setminus \{i,k\}} \sum_{t \in \{i,k\}} (|row_s^*| \cdot |row_t^*|)$ vectors)

$(\chi_u + \chi_v + \chi_w, \chi_{uv} + \chi_{uw} + \chi_{vw})$ $\forall \{v, w\} \in E^*(V^* \setminus T), \exists u \in T$

$\quad (\sum_{s,t \in M \setminus \{i,k\}, s<t} (|row_s^*| \cdot |row_t^*|)$ vectors)

$\square$

In the previous proposition, facets for the $QSAP_{m,N^*}^*$ polytope are generated. We want to gain in addition information, whether there are corresponding facets for the $QSAP_{m,N}$-polytope. The following proposition gives an answer to this question with the help of a zero-lifting technique.

**Lemma 3.24 (Zero-lifting)** *Let*

$$\sum_{v \in V^*} a_v x_v + \sum_{e \in E^*} b_e y_e \leq c \tag{3.9}$$

*be a facet of the $QSAP_{m,N^*}^*$-polytope. The zero-lifting*

$$\sum_{v \in V^*} a_v x_v + \sum_{v \in V \setminus V^*} 0 \cdot x_v + \sum_{e \in E^*} b_e y_e + \sum_{e \in E \setminus E^*} 0 \cdot y_e \leq c \tag{3.10}$$

*defines a facet of the polytope $QSAP_{m,N}$.*

**Proof:**

It is clear that (3.10) is a valid inequality for $QSAP_{m,N}$. For the facet (3.9) of $QSAP_{m,N^*}^*$, we know that there exists a set $P^*$ of $\dim(QSAP_{m,N^*}^*)$ affinely independent vectors in the facet. Since $\phi$ is an isomorphism, the set

$$P := \phi^{-1} \circ \iota(P^*)$$

consists also of $\dim(QSAP_{m,N^*}^*)$ affinely independent vectors. It is easy to see that $P$ is a set of affinely independent vectors in the face of $QSAP_{m,N}$ defined by (3.10). From

$$\dim(QSAP_{m,N}) = \dim(QSAP_{m,N^*}^*)$$
$$= |P^*| = |P|$$

it follows that the face must be a facet. Thus, (3.10) defines a facet of the polytope $QSAP_{m,N}$.

□

The following proposition presents a first facet for the $QSAP_{m,N}$ polytope.

**Proposition 3.25** *The inequality $y_e \geq 0$ defines a facet for the $QSAP_{m,N}$ polytope.*

**Proof:**

We will prove the theorem for different edges of the graph $G$. For $e \in E^*$, we already know from Proposition 3.23 and Lemma 3.24 that $y_e \geq 0$ is facet defining.

Now we consider an edge $e = \{v_{ij}, v_{k1}\} \in E(V^* : V \setminus V^*)$ $(i \neq k)$. From Proposition 3.23, we know that the inequality

$$-x_{ij} + y(v_{ij} : row_k^*) \leq 0$$

defines a facet for $QSAP_{m,N^*}^*$. Combined with the equation

$$-x_{ij} + y(v_{ij} : row_k) = 0$$

of the $(MIP_2)$ formulation, we have

$$y(v_{ij} : row_k) \geq y(v_{ij} : row_k^*)$$

which is equivalent to $y_e \geq 0$. Hence, we get from Lemma 3.24 that $y_e \geq 0$ defines a facet for $QSAP_{m,N}$.

Consider finally for $i \neq k$ the edge

$$e = \{v_{i1}, v_{k1}\} \in E(V \setminus V^* : V \setminus V^*).$$

For the $QSAP_{m,N^*}^*$ polytope, we have the facet defining inequality

$$x(row_i^* \cup row_k^*) - y(row_i^* \cup row_k^*) \leq 1.$$

From the equations

$$
\begin{aligned}
x(row_i) &= 1, \\
x(row_i) &= x(row_i^*) + x_{i1} \text{ and} \\
-x_{ij} + y(v_{ij} : row_k) &= 0
\end{aligned}
$$

we get

$$x(row_i^* \cup row_k^*) - y(row_i^* \cup row_k^*) \leq 1$$

$$\Leftrightarrow x(row_i) - x_{i1} + x(row_k) - x_{k1} - \sum_{v_{ij} \in row_i^*} y(v_{ij} : row_k^*) \leq 1$$

$$\Leftrightarrow 2 - x_{i1} - x_{k1} + y(v_{i1} : row_k) + y(v_{k1} : row_i) - y(v_{i1} : v_{k1}) -$$
$$\sum_{v_{ij} \in row_i} y(v_{ij} : row_k) \leq 1$$

$$\Leftrightarrow 2 - x_{i1} + y(v_{i1} : row_k) - x_{k1} + y(v_{k1} : row_i) - y(v_{i1} : v_{k1}) - 1 \leq 1$$

$$\Leftrightarrow y(v_{i1} : v_{k1}) \geq 0$$

$$\Leftrightarrow y_e \geq 0.$$

With Lemma 3.24 we get that $y_e$ defines a facet for $QSAP_{m,N}$. Thus, we have shown that $y_e \geq 0$ is a facet defining inequality for all $e \in E$.

$\square$

## 3.9.4   Further Facet Defining Inequalities

Saito et al. introduced further facet defining inequalities. These facets are based on two general types of inequalities, namely the clique inequality and the cut inequality. In this subsection, we only present the results. The proofs can be found in the original work ([49]). Since none of the proofs need the sizes $n_i$ of the different rows, the results are also valid for our generalized version of the QSAP.

Both, the clique inequality and the cut inequality are based on a more general inequality that is valid for the $QSAP_{m,N^*}^*$ polytope.

**Proposition 3.26** *For any pair of subsets $S, T \subseteq V^*$ with $S \cap T = \emptyset$ and $\beta \in \mathbb{Z}$, the inequality*

$$-\beta x(S) + (\beta - 1)x(T) - y(S) - y(T) + y(S : T) \leq \frac{\beta(\beta - 1)}{2} \qquad (3.11)$$

*is valid for the $QSAP_{m,N^*}^*$ polytope.*

The following definition introduces the clique inequality and the cut inequality that are both special cases of (3.11).

**Definition 3.27** *By setting $S = \emptyset$ in (3.11), we get the clique inequality*

$$(\beta - 1)x(T) - y(T) \leq \frac{\beta(\beta - 1)}{2} \quad T \subseteq V^*, \beta \in \mathbb{Z}. \tag{3.12}$$

*By inserting $\beta = 1$ in (3.11), we get the cut-inequality*

$$-x(S) - y(S) - y(T) + y(S : T) \leq 0 \quad with \ S, T \subseteq V^*, \ S \cap T = \emptyset. \tag{3.13}$$

For the following theorems, we introduce for $T \subseteq V^*$ the set

$$M(T) := \{i \in M \mid row_i \cap T \neq \emptyset\},$$

which denotes the set of the indices of the rows in which vertices of $T$ occur. For $\beta \leq 0$, the clique inequality (3.12) always holds. Thus, it cannot define a facet. The same holds for any $T \subseteq V^*$ with $|M(T)| = 1$. Here, we have $y(T) = 0$ and thus $x(T) \leq \frac{\beta}{2}$ if $\beta > 1$. The case $\beta < 1$ must not be analyzed, since we have seen before that here the inequality is always fulfilled. From $|M(T)| = 1$, we get $x(T) \leq 1 \leq \frac{\beta}{2}$. Hence, (3.12) is always fulfilled in this case. Thus, for any $T \subseteq V^*$ with $|M(T)| = 1$, (3.12) cannot define a facet of $QSAP^*_{m,N^*}$.

The following two theorems present the requirements of the clique inequality for being facet defining.

**Theorem 3.28** *For $T \subseteq V^*$ with $|M(T)| = 2$, the clique inequality (3.12) defines a facet for $QSAP^*_{m,N^*}$ if and only if*

$$\beta = 1 \ and \ T = \{e\} \subseteq E^*$$

*or*

$$\beta = 2 \ and \ T = (row_i^* \cup row_k^*) \ with \ i, k \in M, \ i \neq k.$$

In the first case, the clique inequality is transformed into the already known facet defining inequality (3.6) ($y_e \geq 0$ for all $e \in E^*$). The second case leads to inequality (3.8), which is

$$x(row_i^*) + x(row_k^*) - y(row_i^* : row_k^*) \leq 1 \quad \forall i, j \in M \ (i \neq k).$$

**Theorem 3.29** *For $T \subseteq V^*$ with $|M(T)| \geq 3$, the clique inequality defines a facet for $QSAP^*_{m,N^*}$ if and only if*

$$2 \leq \beta \leq |M(T)| - 1.$$

The cut inequality cannot be facet defining if $|M(S)| \geq 2$ and $|M(T)|=1$. The following two theorems show special cases of the cut inequality, for which it defines a facet of $QSAP^*_{m,N^*}$.

**Theorem 3.30** *Consider the sets $S, T \subseteq V^*$ with $S \cap T = \emptyset$ and $|M(S)| = |M(T)| = 1$. In this case, the cut inequality (3.13) defines a facet of $QSAP^*_{m,N^*}$ if and only if*

$$M(S) \neq M(T),$$
$$|S| = 1 \ (i.e., \ S = \{v\} \ with \ v \in V^*) \ and$$
$$T = row^*(T)$$

*hold.*

With the requirements of Theorem 3.30, the cut inequality becomes

$$-x_{ij} + y(v_{ij} : row_k) \leq 0 \quad \forall i, k \in M, \ i \neq k, \ \forall j \in N_i^*.$$

For $|M(S)| \geq 1$ and $|M(T)| \geq 2$, we get the following result.

**Theorem 3.31** *Consider $S, T \subseteq V^*$ with $S \cap T = \emptyset$, $|M(S)| \geq 1$ and $|M(T)| \geq 2$. The cut inequality (3.13) defines a facet of $QSAP^*_{m,N^*}$ if and only if*

$\forall \{u_1, u_2\} \in E^*(T) \ \exists v \in S \ s.t. \ \{u_1, u_2, v\} \ is \ a \ 3\text{-clique and}$
$\forall \{v_1, v_2\} \in E^*(S) \ \exists \{u_1, u_2\} \in E^*(T) \ s.t. \ \{v_1, v_2, u_1, u_2\} \ is \ a \ 4\text{-clique.}$

## 3.10 Trivial Bounds for QSAPs

In order to get a first estimation of the optimal solution value of a QSAP, trivial lower bounds can be used. Generating bounds for an optimization problem has the advantage that it is easier to calculate bounds compared to solving a problem exactly. These bounds can be used for an estimation of the solutions of a heuristic approach or to identify room for improvement if an initial solution is given (like the current timetable for the TTSP).

In this section, we introduce different approaches for generating bounds for the QSAP and give some remarks about their usefulness. These bounds have in common that they are easy to calculate, but they differ a lot in the solution quality. More complex bound strategies are presented later in this work in Chapter 5.

### 3.10.1   Row-Pair Bounds

An easy way to compute lower (upper) bounds in $O(m^2n^2)$ is to sum up the minimum (maximum) coefficients for each row pair. For each pair of rows $i, k \in M$ ($i \neq k$), the coefficient $c_{ijkl}$ with the lowest (highest) value is chosen for the bound. With this approach, we get the formulation

$$LB_1 = \sum_{1 \leq i < k \leq m} \min_{\substack{1 \leq j \leq n_i \\ 1 \leq l \leq n_k}} c_{ijkl}$$

for the lower bound. For the upper bound, we have

$$UB_1 = \sum_{1 \leq i < k \leq m} \max_{\substack{1 \leq j \leq n_i \\ 1 \leq l \leq n_k}} c_{ijkl}.$$

### 3.10.2   Triangle Bounds

Better bounds can be achieved in $O(m^3n^3)$ by using row triples. Here, we determine for each row triple the 3-clique with the lowest (highest) sum of the edge weights. Since every row pair is part of $(m-2)$ row triples, we divide the sum of the lowest (highest) 3-clique weights by $(m-2)$ to generate a lower (upper) bound for the problem. Thus, we get for the lower bound

$$LB_2 = \left( \sum_{1 \leq i < k < p \leq m} \min_{1 \leq j \leq n_i} \min_{1 \leq l \leq n_k} \min_{1 \leq q \leq n_p} c_{ijkl} + c_{ijpq} + c_{klpq} \right) / (m-2)$$

and for the upper bound we have

$$UB_2 = \left( \sum_{1 \leq i < k < p \leq m} \max_{1 \leq j \leq n_i} \max_{1 \leq l \leq n_k} \max_{1 \leq q \leq n_p} c_{ijkl} + c_{ijpq} + c_{klpq} \right) / (m-2).$$

### 3.10.3   Star Bounds

Another type of bound that returns promising results in $O(m^2n^2)$ is the star bound. For this bound, the star value is generated for each vertex. For a vertex $v_{ij}$, this value consists of the sum of the lowest (highest) weights of edges to the other rows. Here, one edge is chosen to each set $row_k$ with $k \neq i$. Such a concept of a chosen star for a vertex is presented in Figure 3.8. From each row, the vertex with the lowest (highest) star value is chosen. Since every row-pair is regarded twice in this calculation, the sum of all chosen star values must be divided by two.

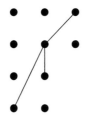

Figure 3.8: Example of the star value for one vertex. The best edges to each row are chosen.

As a formula, the star bound is defined by

$$LB_3 = \left( \sum_{1 \leq i \leq m} \min_{1 \leq j \leq n_i} \sum_{\substack{1 \leq k \leq m \\ k \neq i}} \min_{1 \leq l \leq n_k} c_{ijkl} \right) /2$$

for the lower bound and

$$UB_3 = \left( \sum_{1 \leq i \leq m} \max_{1 \leq j \leq n_i} \sum_{\substack{1 \leq k \leq m \\ k \neq i}} \max_{1 \leq l \leq n_k} c_{ijkl} \right) /2$$

for the upper bound.

### 3.10.4 Pair-Star Bounds

Combining the triangle bound and the star bound leads to a stricter type of bound that we call pair-star bound. This bound can be calculated in $O(m^3 n^3)$. Instead of generating stars for vertices, we construct the star values for edges. For such an edge star, we construct 3-cliques that include the regarded edge and one additional vertex from the other rows. For each of these further rows, we choose the 3-clique with the lowest (highest) edge weights for the edge star. Such a star structure for an edge is shown in Figure 3.9.

For each row pair, we choose the edge star with the lowest (highest) value for the pair-star bound. Note that the chosen edge is counted only once in each star value and not in every 3-clique. In this concept, every row pair is considered more than once. A row pair $(i, k)$ is counted, if

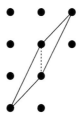

Figure 3.9: Example of the pair-star value for one edge. The best 3-clique is chosen for every additional row.

- - the row pair $(i, k)$ is regarded initially. Here, the row pair is counted once,

- - a row pair $(i, s)$ $(s \neq k)$ is regarded. There are $(m - 2)$ possibilities for $s \in M \setminus \{i, k\}$,

- - a row pair $(t, k)$ $(t \neq i)$ is regarded. There are $(m - 2)$ possibilities for $t \in M \setminus \{i, k\}$.

Thus, every row pair is considered $2 \cdot (m - 2) + 1 = 2m - 3$ times and we must divide the generated sum of the edge star weights by this value. The formulation of the pair-star bound is for the lower bound

$$LB_4 = \Big( \sum_{\substack{1 \leq i < k \leq m}} \min_{\substack{1 \leq j \leq n_i \\ 1 \leq l \leq n_k}} \big( c_{ijkl} + \sum_{\substack{1 \leq p \leq m \\ i \neq p \neq k}} \min_{1 \leq q \leq n_p} (c_{ijpq} + c_{klpq}) \big) \Big) / (2m - 3)$$

and for the upper bound, we have

$$UB_4 = \Big( \sum_{\substack{1 \leq i < k \leq m}} \max_{\substack{1 \leq j \leq n_i \\ 1 \leq l \leq n_k}} \big( c_{ijkl} + \sum_{\substack{1 \leq p \leq m \\ i \neq p \neq k}} \max_{1 \leq q \leq n_p} (c_{ijpq} + c_{klpq}) \big) \Big) / (2m - 3) \,.$$

### 3.10.5    Comparison of the Trivial Bounds

The results of the different trivial lower bound strategies are presented in Section 6.3.1. The solution quality differs a lot for the approaches.

If the different bounds are compared, we can see that the row-pair bound is clearly the weakest and the pair-star bound is the strongest of the approaches. This is plausible, since the computational efforts are the lowest for the row-pair bound and the highest for the pair-star bound. The pair-star bound is able to calculate bounds with gaps lower than five percent for the bigger real life test instances of the TTSP. That is a good result, but it is caused by the special structure of the objective function. For this type of problem, the coefficients $c_{ijkl}$ for fixed $i$ and $k$ are balanced, meaning that they are similar for the different $j \in N_i$ and $l \in N_k$.

If we use the bounds for the test instances that do not have such a structure but for which all coefficients $c_{ijkl}$ are randomly chosen, then the gaps of the lower bounds are larger. For the row-pair bound, gaps of more than 90% make this approach unapplicable for unbalanced problems. Even the better pair-star bound is only able to generate lower bounds that have gaps of more than 50%.

It is noticeable that the gaps of the trivial bound approaches grow larger for bigger problem sizes. Thus, the approaches are not a good choice to generate lower bounds for bigger real life problems. Therefore, we introduce nontrivial lower bound generating approaches that can also deal with larger unbalanced problems in Chapter 5.

## 3.11 Summary of the Chapter

In this chapter, we introduce a generalized version of the QSAP. For this problem, we present an IP, an MIP and a relaxed LP formulation. Furthermore, we present a formulation of the problem on graphs. We show that solving the QSAP is equivalent to finding $m$-cliques in complete multipartite graphs. The complexity of finding the optimal solution of a QSAP is NP-hard. The constraints of the TTSP can be implemented on both, the MIP formulation and the graph formulation.

Saito et al. presented in [49] interesting characteristics of a polytope which describes the QSAP. We extend this result for our generalized version of the QSAP and we prove a new formula for the dimension of this polytope.

We conclude this chapter by presenting first lower bound strategies for the QSAP. These approaches are able to generate lower bounds for small balanced problems, but the gaps of the bounds are very high if applied to larger or unbalanced QSAPs.

# Chapter 4

# Metaheuristic Approaches

The QSAP and the constrained QSAP are both known to be NP-hard. Thus, exact solution strategies cannot solve the problem to optimality in a reasonable amount of time for problems of a larger size. A common approach for generating at least good solutions is to use metaheuristics. Such a heuristic is a problem independent general strategy that is used to generate near optimal solutions. By giving up a guaranteed optimality of the solutions, these algorithms usually find good solutions with acceptable computational efforts.

In this chapter, we introduce three types of metaheuristics that we apply to the QSAP. These metaheuristics are a Simulated Annealing (SA) approach, a Genetic Algorithm (GA) and an Ant Colony Optimization (ACO) algorithm. Since we have a multi-criteria problem, these metaheuristics are enhanced so that they also work for the multiobjective QSAP. We present the multi-colony ACO algorithm in detail, including an adjustment of the parameters. The algorithm is hybridized with a local search approach to improve its efficiency.

Finally, we compare the computational results of the different metaheuristics for a real life test instance.

The goal of this chapter is not to reinvent the wheel or to revolutionize the concept of metaheuristics. Nevertheless, some new aspects and ideas are presented that help to solve the TTSP efficiently. These ideas mainly arise from the multiobjective problem structure.

## 4.1   Literature Survey

NP-hard combinatorial optimization problems have in common that they are hard to solve for bigger problem sizes. Therefore, the usage of metaheuristics for these problems is widely studied. A general overview and a comparison of different metaheuristics for combinatorial optimization problems in the single objective case can be found in [6]. Furthermore, an overview of metaheuristic techniques for multi-criteria combinatorial optimization problems is given in [27].

We apply three different metaheuristics to the TTSP. For the Simulated Annealing approach, the work of Kirkpatrick et al. ([31]) presents a good overview of the different aspects of this type of heuristic. The results for the Genetic Algorithms that we present arise from the master thesis [45], in which Genetic Algorithms are applied to the TTSP. This thesis was done in close cooperation with our work. Furthermore, multiobjective ideas for Genetic Algorithms are presented in [17] and [18].

This chapter presents in detail a multiobjective Ant Colony Optimization approach for the QSAP. A detailed overview of the main aspects of ACO and its applications is presented in [15]. In [14] and [41], the authors analyze the dynamic behavior of ACO algorithms. The Max-Min Ant System approach is presented in [61]. It assures convergence of an ACO algorithm if the pheromone values cannot leave a bounded interval $0 < \tau_{\min} < \tau_{\max} < \infty$. The combination of ACO and constraint satisfaction techniques is presented in [30] and [59]. Finally, some multi-criteria ideas for ACO algorithms are introduced in [37] and [40].

A comparison of these three algorithms and their application to the TTSP that is based on the research of this thesis is presented in [53].

## 4.2   Metaheuristic Approaches for the QSAP

Solving QSAPs is, as we have seen, not an easy task. The large amount of local minima (cf. Figure 2.9) makes it difficult for the metaheuristics to find global good solutions, since they can easily get trapped in a local optimum. The same problem also occurs for multi-criteria problems. Here, the goal is not only to find a single optimum, instead we are searching for a set of Pareto optimal solutions.

We differ in this work between modifying algorithms and constructive algorithms. A modifying algorithm starts with one or several given initial solutions and tries to improve them by certain strategies (SA or GA are examples for modifying algorithms). Conversely, a constructive algorithm generates solutions from scratch by using knowledge gained in the earlier phase of the algorithm (ACO is an example of a constructive algorithm).

## 4.2.1 Simulated Annealing

Simulated Annealing is a metaheuristic, whose name and inspiration come from annealing in metallurgy. Here, heating and controlled cooling of a material is used to increase the size of the crystals and to reduce the defects in the material.

A high temperature represents the ability of the algorithm to leave a local minimum in which it got stuck. A controlled cooling reduces the temperature over time so that the algorithm finally ends in a local good solution.

Given a start solution, the algorithm successively tries to find a better solution in its neighborhood. The temperature of the process is converted in the SA approach into a probability that allows small worsenings of the current solutions while choosing a solution from its neighborhood. This strategy prevents the algorithm from getting stuck in a local optimum in its early phase. The controlled cooling reduces the temperature over time. Thus, the percentage that allows worsenings in the algorithm is also lowered. Each single run of the algorithm ends if the temperature reaches the value zero.

Since we regard a multi-criteria problem, the choice of a neighborhood and the evaluation of the solution quality needs special handling. To deal with the higher dimensional objective space, we let the algorithm run several times for a high number of start solutions. For each of these solutions, a different "direction" in the search space is chosen. This is done by using a weighted sum of the normalized objective function values as the "energy" of a solution. Thus, in each step of a run, we choose the solution in the neighborhood with the lowest energy. The weights are generated so that all areas of the three-dimensional Pareto frontier are sufficiently analyzed. The number of simultaneously analyzed directions is increased during the runtime of the SA. The new directions that are added are not necessarily chosen homogeneously, but rather with the goal to achieve a more precise approximation in promising areas of the approximated Pareto frontier.

A solution of a QSAP is an $m$-tuple, where each element represents the choice of a location $j \in N_i$ for an object $i \in M$. For an improvement step, we generate the regarded neighborhood by selecting a random object. From the possible locations for this object, we pick a random subset of $p$ percent. This percentage that defines the size of the neighborhood is increased during the runtime. This is not typical for SA approaches, but the test runs show that this strategy further improves the efficiency of the algorithm. From the set of new possible solutions, we choose the neighbor with the lowest energy (the smallest weighted sum of the normalized objective function values). If this new solution is better than the old one, the algorithm proceeds with this solution. If the solution is worse, the new solution is only chosen with a probability that depends on the current temperature of the algorithm. This is done until the cooling process is finished and the algorithm has found a hopefully good solution.

By repeating this process for different start solutions and with different search directions in the objective space, a good approximation of the Pareto frontier can be achieved for the QSAP. More details of our SA approach for the TTSP are presented in [53].

## 4.2.2  Genetic Algorithm

A Genetic Algorithm is a technique that is inspired by biological evolution. It uses evolutionary ideas such as inheritance, mutation, selection and crossover. A start population of solutions is generated and combined pairwise to create a new set of offspring solutions. This crossover process tries to combine the good characteristics of two parents to generate promising offspring. The inheritance is followed by a mutation, where the offspring solutions are slightly modified. The most promising offspring solutions according to a fitness function are selected for the next evolutionary step. A GA is, like the SA, a modifying algorithm, since a set of starting solutions is required.

The GA approach for the TTSP is presented in detail in [45]. Here, the author encodes the changes that are made to a timetable by a chromosome representation. Each chromosome consists of a set of genes and each gene corresponds to a shift for a specific line. For the crossover of two chromosomes, a randomly chosen split point defines the crossover strategy of two parents. One parent contributes the genes of the first part and the other parent contributes the genes of the second part of the chromosome. The mutation of the child solution is done by randomly changing one gene to a different value.

The algorithm starts with a randomly generated initial population. Alternatively, one can also give the algorithm a warm start by providing a set of precomputed good solutions.

### 4.2.3 Ant Colony Optimization

Ant Colony Optimization is a modern metaheuristic that copies the behavior of ant colonies that are searching for food. Here, a high number of individual ants swarm out to search in the local neighborhood of the anthill for promising food sources. The ants share information via pheromone which they place on their routes. After a short period of time, the colony is not only able to inspect a big area, but also to find the shortest paths to the food sources. This self-organizing process of the swarm can be used for optimization purposes. Here, several agents copy the behavior of the ants and the following agents benefit from the knowledge of previous agents by analyzing the pheromone trails.

We introduce our approach of using ACO algorithms for the QSAP in the following section. Here, we also discuss different theoretical and practical aspects of this metaheuristic. To deal with the multi-criteria structure of our problem, several ant colonies work on different parts of the objective space simultaneously.

## 4.3   Ant Colony Optimization

Ant colony algorithms have been researched intensively in the last fifteen years. Most of the work focuses on single-objective problems, but recently more attention has been paid to multiobjective optimization as well. Here, several ant colonies are used to deal with the multi-criteria aspects of the problem.

In this work, we combine our multi-colony approach with some techniques to solve the constrained QSAP. The constraints are integrated in our model by removing vertices and edges from the construction graph that the single ants use to generate solutions. Here, the ants try to find good solutions while considering all constraints.

Some similar problems, like Quadratic Assignment Problems (QAP) and Constraint Satisfaction Problems (CSP), have been solved with ACO techniques quite successfully. Thus, using ACO for the constrained QSAP is promising.

### 4.3.1   Introduction to ACO

The ACO algorithm imitates the behavior of real ants and uses the stigmergy effect, meaning that ants communicate indirectly through pheromone trails which they place on their routes. An ant that has found a promising food source returns to the anthill and influences the following ants by depositing pheromone on its route. A first impression of the strategies of ants, their behavior and the self-organizing process they create can be found in [20].

The classical ACO algorithm uses a single colony of ants as virtual agents to scan the search space for feasible good solutions until a stopping criterion or a fixed number of generations of ants is reached. All ants of a generation start their search simultaneously, meaning that they use the same pheromone information. By marking good solutions with a higher amount of pheromone, the ants communicate with each other. Thus, the following generations can use this improved knowledge of the problem structure to find better solutions. Over time, the pheromone trails evaporate with a constant factor $(1 - \rho)$ $(0 < \rho < 1)$. Thus, older solutions that are dominated by new solutions have less influence on future ants. An overview of the different aspects of ACO algorithms is presented in [15].

We now sketch the basic ideas of the algorithm. On a problem specific construction graph, each ant starts at a randomly chosen vertex and selects from the outgoing edges stochastically a promising one (i.e., one with a high pheromone value). Edge by edge, the ant continues to construct a solution by using the information that previous ants left. Each constructed solution is compared to the set of the overall best found solutions and the solutions found by ants of the same generation. If the solution is promising, the ant places a certain amount of pheromone on all edges that belong to the corresponding solution.

To avoid stagnation, the pheromone trails are reset to their start values if the algorithm does not find new good solutions in a certain number of generations. These resets give the algorithm a higher exploration ability and prevent it from being stuck in a local optimum.

An approach that ensures that the algorithm converges is to use the Max-Min Ant System ($\mathcal{MMAS}$) concept from [61]. In this approach, a minimum and a maximum pheromone value ($\tau_{\min}$, $\tau_{\max}$) are defined. These bounds cause a larger exploration of the algorithm, since they avoid that choices are not considered anymore due to a low pheromone value. In addition, the upper bound prevents that a certain choice dominates all other possibilities if the corresponding pheromone value becomes too high. Thus, the algorithm converges in value which means that it is able to generate all possible solutions if the runtime is long enough. The values $\tau_{\min}$ and $\tau_{\max}$ must be adapted to the specific optimization problem.

All pheromone values are initialized at the beginning with the pheromone value $\tau_{\max}$. In case of a reset of the pheromone values, we also set all values to $\tau_{\max}$ to prevent stagnation. We choose the maximum value to guarantee a higher exploration rate in the first iterations of the optimization process.

Another way to influence the exploration rate is to use the pseudorandom proportional rule. Here, an exploration rate $q_0 \in [0, 1]$ is introduced. Each time an ant has to decide which edge to take next, it uses the stochastic approach with probability $q_0$. Otherwise, it takes the absolute best choice (in terms of the corresponding pheromone values) with probability $(1 - q_0)$. The exploration rate regulates the degree of exploration by influencing the choice whether to concentrate the search close to the best-so-far routes or to explore new routes. An analysis of good choices for the evaporation rate $\rho$ and the exploration rate $q_0$ is presented in Section 4.4.1.

## 4.3.2   ACO for Semi-Assignment Problems

We use the QSAP-graph $G_{QSAP} = (V, E)$ from Section 3.6.2 as the construction graph. This graph is presented in Figure 4.1. The graph $G_{QSAP}$ is a multipartite graph and the vertices can be partitioned into the sets $\mathcal{V} = \{V_1, ..., V_m\}$ ($V_i \subseteq \mathcal{V}$). The ants generate the solutions ($m$- cliques) on this graph, the pheromones are placed on the edges. Here, the pheromone value $\tau_{ijkl} = \tau_{klij}$ is the amount of pheromone on edge $e_{ijkl}$ ($i < k$). A higher pheromone value on an edge indicates that it is promising to choose both vertices for a solution.

We construct a solution for an ant as follows: The ant starts in a uniformly chosen subset $V_i \in \mathcal{V}$. We choose such a subset to reduce the computational effort that would occur from comparing all vertices (cf. [59]). In this way, we

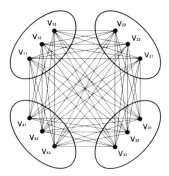

Figure 4.1: Example of a construction graph $G = (V, E)$.

just have to examine the pheromone trails connected to vertices contained in this set $V_i$. The vertex $v_{ij} \in V_i$ at which the ant starts is computed stochastically. For this decision, heuristic information $\eta_{ij}$ and weights $w_{ij}$ that are calculated from the pheromone values $\tau_{ijkl}$ are taken into account.

The probability that an ant chooses a certain vertex $v_{ij}$ from the start set $V_i$ is

$$p_{ij} = \frac{(w_{ij})^\alpha (\eta_{ij})^\beta}{\sum_{k=1}^{n_i} (w_{ik})^\alpha (\eta_{ik})^\beta} \ .$$

The value $\eta_{ij}$ represents the problem dependent heuristic information that vertex $v_{ij}$ should be chosen. The parameters $\alpha$ and $\beta$ satisfy $\alpha + \beta = 1$ and they control how much the heuristic information and the pheromone values influence the probability $p_{ij}$. For the TTSP, the value $\eta_{ij}$ refers to a preference of a certain shift of a line. But such a preference is not given in our problem, since the quality measure of the solutions only depends on pairs of shifts and not on single shifts. Thus, we set $\beta = 0$ and $\alpha = 1$. This is not an uncommon approach. An advantage of ACO is that the algorithm is able to generate good solutions even if no heuristic information is given. This results in a longer adjustment time at the beginning of the run of the algorithm, but it also leads to a better covering of the search space.

The weights $w_{ij}$ are calculated differently for the first vertex set compared to the other vertex sets. For the vertices from the first vertex set $V_i$, we define the weights $w_{ij}$ by

$$w_{ij} = \sum_{\substack{k=1 \\ k \neq i}}^{m} \sum_{l=1}^{n_k} \tau_{ijkl} \quad \forall j \in \{1, ..., n_i\},$$

where $\tau_{ijkl}$ is the amount of pheromone on the edge between vertex $v_{ij}$ and vertex $v_{kl}$.

After selecting the first vertex, the algorithm chooses the next vertex not by looking at all pheromone values on the edges that are incident to that new vertex. Instead, we just examine the pheromone trails on edges that are incident to already visited vertices. This is a difference to most other ACO algorithms, where the choice of the next vertex only depends on the local information of the last visited vertex. In our approach, the decision of the algorithm is based on the global information of all visited vertices of an ant (cf. [59]).

After the first vertex is chosen, the algorithm selects again a random subset $V_i \in \mathcal{V}$ that the ant has not visited yet. For all vertices $v_{ij}$ from this subset, the weights $w_{ij}$ are calculated as follows: Let

$$P = \{\langle \lambda_1, \mu_1 \rangle, ..., \langle \lambda_p, \mu_p \rangle\}$$

be the path through the vertices that the ant has already visited. Here, the algorithm has chosen vertex $v_{\lambda_s \mu_s}$ from subset $V_{\lambda_s} \in \mathcal{V}$. For the next choice, we get the weights

$$w_{ij} = \sum_{k=1}^{p} \tau_{ij\lambda_k\mu_k} \quad \forall j \in \{1, ..., n_i\}$$

which are used to calculate the probabilities $p_{ij}$.

Another difference between our approach of the ACO for the Semi-Assignment Problem and the standard ACO techniques is that the ants do not only place pheromone on their chosen path through the graph, but on all edges that belong to the clique induced by the visited vertices. This is due to the fact that a good assignment results from good shift pairs for all line pairs. The order in which the vertices are visited is not relevant for the given problem.

### 4.3.3 A Multi-Colony Approach

Since we have several optimization criteria in our approach, it does not suffice to use just one pheromone trail to store the different types of information that the ants have to share. If all non-dominated ants would provide their pheromone to a single trail, the "noise" would be disturbing and it would be impossible to extract useful information from the trails. Ants for which the algorithm tries to find good solutions for one criterion would also follow the trails from ants that found good solutions for another criterion.

That is why we use a multi-colony approach, where different colonies are responsible for approximating a specific area of the Pareto frontier (cf. [40]). For this task, each colony has its own pheromone trail. During the run of the algorithm, the set of solutions that are not dominated change and the colonies adapt their specific areas. Every ant that starts to construct a new solution belongs to a single colony and reacts only to the corresponding pheromone trail.

We store all non-dominated solutions that are found by ants. When all ants of a generation have finished their search, the new solutions are compared to the set of the non-dominated solutions. If new non-dominated solutions are found, we add them to this set. Furthermore, if a formerly non-dominated solution becomes dominated, we remove it from this set. Only the ants whose solutions belong to this set add information to the pheromone trails. This is similar to the global best approach from the one-dimensional ACO, where not the best ants of the last generation but the best ants of all generations are allowed to update pheromone values. Furthermore, the ants do not necessarily add pheromone to the trail that they used to construct their solution. Instead, they place pheromone of the type that corresponds to the colony of the area where their solution is located.

There are several ways to divide a Pareto frontier into disjoint areas for the colonies. We present here an idea for two and for three objective functions.

An easy way to divide a two-dimensional Pareto frontier is by cutting the area in three parts with a similar number of non-dominated solutions. An example of such a partition can be seen in Figure 4.2. Here, we sort all solutions with respect to the result for the first objective function. Then we divide the solutions in sets of a similar size ($\pm 1$). Each of these sets is assigned to an ant colony.

If there are three objective functions, one can make the partitioning of the

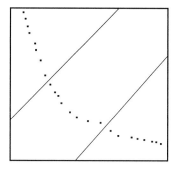

Figure 4.2: Example of a partitioning of a two-dimensional approximated Pareto frontier.

objective space in the following way: We begin by dividing the approximated Pareto frontier into several areas by sorting the non-dominated solutions according to their results for the first objective function. We split the results into three equally sized ($\pm 1$) sets $R_1$, $R_2$ and $R_3$. We sort each of these sets according to the second optimization criterion and split the sets $R_i$ ($i \in \{1, ..., 3\}$) again in three equally sized ($\pm 1$) sets $R_{i1}$, $R_{i2}$ and $R_{i3}$. These nine rectangles contain the corresponding solutions inside a cuboid with the third objective function as the third dimension. We divide these cuboids again into three smaller cuboids so that the same number of solutions ($\pm 1$) can be found in all of them. With this approach, we get 27 ant colony areas that cover the objective space. This amount of colonies is still manageable by modern computers for problem sizes that correspond to real life problems.

An example of a partitioning for a three-dimensional solution set is shown in Figure 4.3.

The partitions are not fixed for the whole optimization process, since the set of the non-dominated solutions change during the run of the algorithm. Thus, the areas that are optimized by the different colonies also change with the partitions. Hence, the same ant can add pheromone to different pheromone trails during the run of the algorithm, if the areas of the colonies change so that the solution belongs to a different colony.

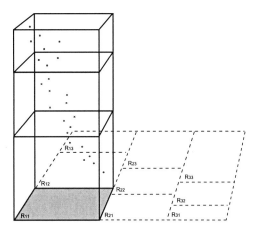

Figure 4.3: Partitioning example of a three-dimensional approximated Pareto frontier.

### 4.3.4   Constraint Handling

This subsection deals with the problem of using constraints in the ACO algorithm. The types of constraints that we have to deal with for the TTSP are presented in Section 2.7. These constraints forbid certain shifts for lines or certain shift combinations for line pairs. In the first case, we remove the corresponding vertices and the adjacent edges from the construction graph. In the second case, we can remove the edges that belong to these shift combinations.

In the presence of the second type of constraints, $G$ is no longer a complete multipartite graph (cf. Figure 4.4). However, feasible solutions still must correspond to $m$-cliques in $G$. These cliques must contain one vertex from each set $V_i$.

We use the same ACO algorithm for the constrained QSAP that we introduced for the unconstrained problem. Here, the algorithm has to ensure that the next vertex that is chosen for an ant still satisfies all constraint. This means that all edges of the induced clique are contained in $G$. We use again the transition probabilities

$$p_{ij} = \frac{(w_{ij})^\alpha (\eta_{ij})^\beta}{\sum_{k=1}^{n_i} (w_{ik})^\alpha (\eta_{ik})^\beta}$$

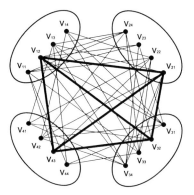

Figure 4.4: Construction graph $G = (V, E)$ with constraints and a feasible assignment (marked by the bold lines).

from Section 4.3.2. But with constraints, we need a modified formulation of the weights $w_{ij}$. To ensure that no constraints are violated, we introduce a new variable $\delta_{ijkl}$ with

$$\delta_{ijkl} = \left\{ \begin{array}{l} 1, \text{ if } \{v_{ij}, v_{kl}\} \in E, \\ 0, \text{ else.} \end{array} \right.$$

With these variables, we define the weights $w_{ij}$ as follows: For the first vertex that the algorithm chooses, we use

$$w_{ij} = \sum_{\substack{k \in \{1,\dots,m\} \\ k \neq i}} \sum_{l=1}^{n_k} \tau_{ijkl} \cdot \delta_{ijkl} \ .$$

Afterwards, if an ant has already constructed a path

$$P = \{\langle \lambda_1, \mu_1 \rangle, \dots, \langle \lambda_p, \mu_p \rangle\}$$

with at least one vertex, we use the weights

$$w_{ij} = \sum_{k=1}^{p} (\tau_{ij\lambda_k\mu_k}) \cdot \prod_{k=1}^{p} (\delta_{ij\lambda_k\mu_k}) \ .$$

The product of the $\delta$ variables is used, since a vertex should not be chosen if any of the $\delta$ variables has value zero. Thus, the weight has value zero if the choice of an edge would result in a path that induces no clique in $G$.

If constraints are applied to the problem, attention must be paid whether there is a feasible way for an ant to continue its path. If all the weights $w_{ij}$ that belong to a chosen set $V_i$ have value zero, i.e.,

$$\sum_{j=1}^{n_i} w_{ij} = 0,$$

the ant has found a dead-end route and will stop searching for further vertices to create a solution (we do not calculate $p_{ij}$ in this case, since the denominator would have value zero). This guarantees that, at any time, the vertices of a chosen path $P$ induce a clique in $G$.

An advantage of the use of an ACO algorithm for solving constrained problems is that, after several ant generations, the knowledge about feasible solutions is integrated in the pheromone trails. Edges that belong to no or just a few feasible solutions get less pheromone during the run of the algorithm than edges that belong to many solutions. Compared with metaheuristics that construct solutions by changing initial solutions, this is clearly an advantage. Especially in highly constrained search spaces, the knowledge of previous ants can be very helpful.

### 4.3.5    Structure of the ACO algorithm

The pseudocode of the Constraint-Multi-Colony-ACO algorithm is presented in Figure 4.5. Not all operations of the ACO algorithm are directly related to single ants. These additional actions that influence all pheromone values at the same time, e.g., pheromone evaporation or a reset of a pheromone trail, are called daemon actions.

The ants, whose solutions are non-dominated, add a constant amount of pheromone to the trails during the pheromone update phase. The amount of pheromone depends on the choice of the values $\tau_{\min}$ and $\tau_{\max}$ and the number of non-dominated solutions in each colony. For both, the pheromone evaporation and the pheromone update steps, we have to make sure that the pheromone values do not leave the allowed interval $[\tau_{\min}, \tau_{\max}]$.

### 4.3.6    Convergence Theorem

The following theorem shows that our algorithm has convergence in value. It is based on the ideas of a theorem from [15]. Convergence in value means

---

**Algorithm 4.3.1:** CONSTRAINTMULTICOLONYACO$(\rho, q_0, \alpha, \beta, \kappa)$

Select constraints
Initialize pheromone trails to $\tau_{\max}$
**for** $i \leftarrow 0$ **to** number of generations
  **do**
    **for** $j \leftarrow 0$ **to** number of colonies
      **do**
        **for** $k \leftarrow 0$ **to** number of ants per colony
          **do**
            Create a new ant
            Construct a solution for this ant
            Compare the solution with the non-dominated solutions
      DaemonAction:   Pheromone evaporation with rate $\rho$
      Update pheromone trails for non-dominated solutions
      DaemonAction:   Reset pheromone trails of colonies for which
        no non-dominated solutions were found in the last
        $\kappa$ generations to $\tau_{\max}$

---

Figure 4.5: Pseudocode of the Constraint-Multi-Colony-ACO algorithm.

that the algorithm is able to find all non-dominated solutions if the algorithm runs long enough.

**Theorem 4.1** *Let $\mathcal{A}_f$ be the nonempty set of all feasible assignments and $A$ be one of these assignments $A \in \mathcal{A}_f$ with all $\eta_{ij} > 0$ for this assignment $A$. Let $p_A(n)$ be the probability that Algorithm 4.3.1 with $\alpha > 0$ finds the assignment $A$ within the first $n$ iterations. Then, for an arbitrary small $\varepsilon > 0$, there exists a sufficiently large $n_0$ for which*

$$p_A(n_0) \geq 1 - \varepsilon$$

*holds. This leads asymptotically to*

$$\lim_{n \to \infty} p_A(n) = 1.$$

**Proof:**

The assignment $A$ corresponds to a clique $C$ in the graph $G_{QSAP}$. Thus, all edges from this clique must exist in the graph $G$. Hence, we have

$$\delta_{ijkl} = 1$$

for all binary variables $\delta_{ijkl}$ that correspond to edges between vertices $v_{ij}, v_{kl} \in C$.

Let $p_0$ be the smallest probability that can occur during the algorithm that an ant chooses one vertex of the clique as a starting point. Note that the probabilities $p_{ij}$ depend on the current pheromone states. Since we set limits $0 < \tau_{\min} < \tau_{\max} < \infty$ for the pheromone values, the numerator of all possible $p_{ij}$ is greater than zero and it follows directly that $p_0 > 0$ holds.

After the algorithm has chosen at least one vertex for an ant, we can generate for every vertex $v_{ij}$ of a new subset $V_i \in \mathcal{V}$ the probability $p_{ij} > 0$ that the algorithm chooses $v_{ij} \in V_i$ as the next vertex for an ant. We define the probability $p_{\min}$ as the smallest possible of these probabilities that can occur during the run of the algorithm. Since the pheromone values are bounded by $\tau_{\min}$ and $\tau_{\max}$, it follows again that $p_{\min} > 0$ holds.

The probability $p_A$ that assignment $A$ is generated by an ant satisfies

$$p_A \geq (p_{\min})^{m-1} \cdot p_0.$$

Now we consider the probability $p_A(n)$ that $A$ is generated by at least one ant out of $n$ ants. Thus, we get from the inequality

$$1 - p_A(n) \leq (1 - (p_{\min})^{m-1} \cdot p_0)^n$$

an upper bound for the probability that $n$ ants do not generate assignment $A$. This can be transformed into

$$p_A(n) \geq 1 - (1 - (p_{\min})^{m-1} \cdot p_0)^n.$$

This inequality directly implies that the probability $P_A(n)$ becomes larger than any $(1 - \varepsilon)$ for a sufficiently large $n$. It follows that

$$\lim_{n \to \infty} p_A(n) = 1$$

holds.

$\square$

A problem of the result of this theorem is that we do not know how fast the algorithm converges. We also have no estimation of the time that is needed for finding at least a good approximation of the Pareto frontier. Some work on this problem was done for single-objective optimization in [14].

# 4.4 Computational Results

In this subsection, some computational results for real life case studies are presented. But before we start the optimization process, we have to adjust our ACO parameters to the QSAP of the TTSP. Therefore, we make some tests to enhance these parameter values for the one-dimensional TTSP for the city of Kaiserslautern with the objective function $\psi_1$ and without constraints.

## 4.4.1 Parameter Adjustment

Not much is known for good choices of the values $\tau_{\min}$ and $\tau_{\max}$, since these parameters must be chosen problem dependent. Thus, general recommendations for the values are not possible. In our work, we choose $\tau_{\min} = 20$ and $\tau_{\max} = 100$ after some tests to keep the exploration of the algorithm high.

Recent research (cf. [15]) showed that small values for the evaporation rate $\rho$ provide good results for the Max-Min concept, so we tested our algorithm with values $\rho \in \{0.02,\ 0.05,\ 0.1\}$. The parameter $\rho$ indicates how much pheromone evaporates after each generation. Thus, the evaporation rate regulates the influence of older and dominated solutions on the future progress of the algorithm. The average results of several runs for different test instances are presented in Figure 4.6. The best results are achieved with $\rho = 0.05$. Hence, we use this evaporation parameter for our analysis.

Another important parameter is the exploration rate $q_0$ that regulates the influence of the pseudorandom proportional rule. This value influences the algorithm to keep the ants close to already found good solutions or to explore new areas of the search space. Appropriate values for $q_0$ are unknown for the QSAP, so we tried the values $q_0 \in \{0.1,\ 0.2,\ 0.5\}$. Figure 4.7 shows the average results for the different values.

The smaller values $q_0 = 0.1$ and $q_0 = 0.2$ achieved better results, so we take $q_0 = 0.2$ for our ACO algorithm, since this parameter setting performs better in the early phase of the test runs. As expected, the optimization

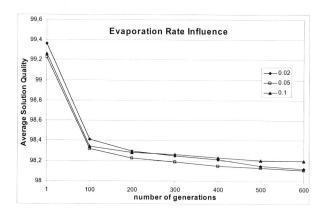

Figure 4.6: Average solution quality during the optimization process for different evaporation rates. The $y$-axis represents the percentage improvement of the timetables ($\psi_1$).

runs with a larger value start with better results (because of the higher exploration), but in the later phase of the algorithm, focusing on already found good areas performs better.

## 4.4.2   Optimization Results

The following analysis examines the behavior of the algorithm for a multi-objective problem. We consider the objective functions $\psi_1$, $\psi_2$ and $\psi_3$ from Section 2.5 for the QSAP that arises from a real life application of the TTSP.

In Figure 4.8, we present a two-dimensional projection of the three-dimensional approximated Pareto frontier. The $x$-axis presents the percentage improvement of the objective function $\psi_1$ and the $y$-axis shows $\psi_3$, the amount of changes that are made to the timetable. We compare the results of an optimization with constraints (gray) to the results of an unconstrained optimization (black). Every point in the figure represents a non-dominated solution. It can be seen that the approximated Pareto frontier from the unconstrained problem is destroyed if constraints are added to the problem. Without constraints, improvements of up to four percent can be achieved for $\psi_1$. When the constraints are added, the best achieved results have an improvement of about one percent.

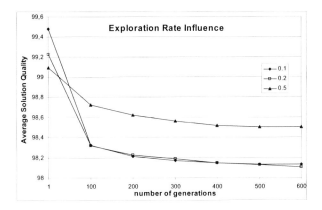

Figure 4.7: Average influence of the exploration rate on the optimization process. The $y$-axis represents the percentage improvement of the timetables $(\psi_1)$. A higher exploration rate results in a better solution quality in the early phase of the algorithm. For longer optimization runs, a smaller exploration rate leads to a better solution quality.

An interesting aspect of the ACO approach is that the constraints can be changed during the runtime of the algorithm. The ants will react to the different environment and, after some time, the pheromone trails adapt to the changes so that they reflect the new problem setting properly.

An analysis of the solution quality of the ACO algorithm for the TTSP is presented in Chapter 6 and in Chapter 7. In Section 6.5, we compare the approximated Pareto frontier that is generated by the ACO algorithm to exact solution strategies. It can be seen that the approximated Pareto frontier of the ACO approach gets very close to the optimal Pareto frontier, even for large real life test instances. The analyzed results are computed by an approach, in which the ACO algorithm is hybridized with a local search approach. This hybridization is presented in the following section.

## 4.5 Hybridization with Local Search

For many combinatorial optimization problems that are solved with meta-heuristics, a hybridization with a local search approach shows promising re-

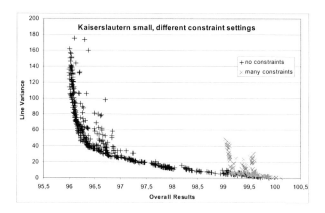

Figure 4.8: Comparison between a constrained and an unconstrained optimization for the small Kaiserslautern case study.

sults. A local search algorithm takes an initial solution and searches in its local neighborhood for improvement possibilities. In most cases, the metaheuristic is used to generate initial solutions and the local search process further improves the generated solutions. In [15], the authors propose to add a local search step to an ACO algorithm to increase the efficiency of the algorithm. Further research was done on this topic in [60], where the author compares the effects of different local search approaches for various optimization problems.

We also integrate such a local search step in our metaheuristics for the TTSP. For a generated solution, the sets $V_i$ are analyzed consecutively for possible improvements via changes of the chosen vertex $v_{ij}$. Here, a vertex $v_{ij'}$ is preferred to the initial vertex $v_{ij}$ of a solution, if the new generated solution has a better value for a weighted sum of the objective functions. Such a sum is given in the form

$$\psi(x) = \alpha_1 \cdot \psi_1(x) + \alpha_2 \cdot \psi_2(x) + \alpha_3 \cdot \psi_3(x).$$

The weights $\alpha_1$, $\alpha_2$ and $\alpha_3$ are chosen for an initial solution according to its position with respect to the approximated Pareto frontier. This choice increases the chance to find new non-dominated solutions. We continue the process of improving a certain solution until no improvements can be achieved for all sets $V_i$ ($i \in M$). This means that a local optimum is reached with respect to the weighted sum.

The solution quality of the hybrid algorithm is significantly better than the results of the algorithm without a hybridization. Especially the speed of convergence is much faster. This affirms the general result from other papers that an additional local search step improves the efficiency of a metaheuristic.

A comparison of the results of the multi-colony ACO without a local search process and the hybrid version that contains a local search step is presented in Figure 4.9 for a large real life test instance. Here, the approximated Pareto frontier that is calculated after ten minutes is shown. It can be seen that the hybrid approach shows a much better performance.

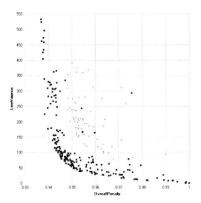

Figure 4.9: Comparison of the ACO results for a large real life test instance with local search (black points) and without local search (gray points) after 10 minutes of computation time. The $x$-axis represents the percentage improvement of the overall solution quality $\psi_1$ and the $y$-axis shows the line variance $\psi_3$.

## 4.6 Comparison of the Approaches

In this section, we compare the results of the three metaheuristics SA, GA and ACO. A more detailed version of our results can be found in [53]. The algorithms are compared for a real life test instance for the city of Kaiserslautern. We analyze a TTSP with three objective functions ($\psi_1$, $\psi_2$ and $\psi_3$). The test instance corresponds to a QSAP with $m = 31$ and different sizes of the sets $N_i$. The values $n_i$ are between 1 and 30.

A modern MIP solver (Cplex 11.2) needs slightly more than 25 hours to solve the problem exactly with only a single objective function ($\psi_1$). For the metaheuristics, we approximate the Pareto frontier 30 times with each algorithm and compute the relative volumes of the approximated Pareto frontiers after three time milestones. These milestones are represented by the number of evaluations of the objective functions, which is by far the most time consuming part of the calculations. These milestones correspond roughly to times of about one minute, ten minutes and one hour. The relative volume is calculated as follows: We start by generating the set of all non-dominated solutions that are found by all runs of the three algorithms. In addition, we need a point of reference that is in all three criteria worse than the worst solution. With the three dimensional set of solutions and the reference point, we can compute a volume. This volume is then used to evaluate the results of the different approaches.

The relative volume of the result of a single run of an algorithm is the percentage of the volume of the set of the non-dominated solutions of the single run from the volume of the set of the non-dominated solutions of all runs. For further details on this concept for comparing different multi-dimensional solution sets, see [53].

The computational results are presented in Figure 4.10. They show the promising average Pareto frontier volumes of the different approaches at the three time milestones. It can be seen that the results of the SA and the ACO approach have a relatively stable convergence behavior, represented by the top and bottom lines. Note that the SA and the ACO algorithm both work with a hybridization with a local search algorithm. The solutions of the GA approach were not improved with such a strategy. This explains the significant difference of the results.

It can be seen that both, the SA and the ACO approach, perform good. Promising results can be achieved in a short time, which is also a consequence of the hybridization with local search. In Section 6.5, we present a comparison of the ACO results to the exact Pareto frontier. It can be seen that the approach is able to approximate the set of the Pareto optimal solutions quite well. Thus, the TTSP can be efficiently solved with the presented

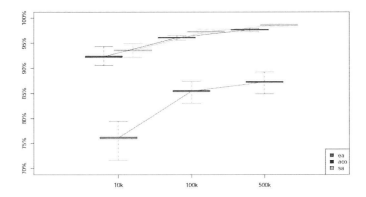

Figure 4.10: Results that show how close the solution sets of the metaheuristics get to the total best approximation of the Pareto frontier for three time milestones. The $y$-axis presents the average relative volume of the different approaches. The gray curve at the bottom presents the results of the Genetic Algorithm, the black curve shows the ACO approach and the light gray curve shows the Simulated Annealing method. The thick horizontal lines denote the means and the top and bottom lines denote the maximum and the minimum result.

metaheuristics.

## 4.7   Summary of the Chapter

In this chapter, we introduce three different metaheuristics that are able to find good solutions for the multiobjective QSAP. These heuristics are a Simulated Annealing approach, a Genetic Algorithm and an Ant Colony Optimization algorithm. We present the multi-colony ACO approach in detail, including a scaling of the parameters for the TTSP.

In addition, a hybridization of the approaches with a local search method is proposed. This combination of different heuristics provides promising results for the QSAP. Finally, a comparison of the three different metaheuristics is presented.

# Chapter 5

# The Reformulation Linearization Technique for the QSAP

In the previous chapter, solutions for the QSAP were generated by metaheuristics. But the quality of these solutions is unknown and we need a method to evaluate the solutions. This problem can be solved by calculating lower bounds for the problem. One method for generating bounds for QAPs is the Reformulation Linearization Technique (RLT) from [56], which provides good lower bounds in a reasonable amount of time. Since the QSAP has a quite similar structure to the QAP, we introduce RLT variants for the QSAP in this chapter.

The RLT provides a hierarchy of stepwise tightening polyhedral formulations of the problem. We analyze these formulations for the QSAP and we are able to determine the minimal RLT level that is needed for an exact characterization of the original QSAP. This level was up to now unknown. In addition, we present a new RLT-1.5 algorithm that successively eliminates a structure that is responsible for untight solutions.

The RLT has a strong connection to the concept of the height from the research area of pseudo-Boolean functions. The height is the maximal constant value that can be extracted from the objective function of a QSAP. By approximating this height with a Dual Ascent method, we get a new strategy to generate lower bounds for the QSAP. Finally, we present a hybrid algorithm that combines the convergence speed of the Dual Ascent method with the strengths of the RLT-1.5 approach.

## 5.1 Introduction of the RLT

The RLT is a method for generating tight linear or convex relaxations for discrete and continuous nonconvex programming problems. It works in two phases. In the first phase, a reformulation step is made which generates new nonlinear valid inequalities. The second phase linearizes the product terms that are generated in the reformulation step by continuous variables.

The technique builds an $n$ level hierarchy, where a finite number $n$ provides an RLT-level-$n$ (or RLT-$n$) formulation that characterizes the convex hull of the solutions of the initial problem.

The projection of the higher dimensional polyhedra to the space of the original $x$-variables and $y$-variables of the LP formulation of the QSAP (cf. Section 3.4.2) provides stepwise improving polyhedral formulations

$$X_{P_0} \supseteq X_{P_1} \supseteq X_{P_2} \supseteq ... \supseteq X_{P_n} = conv(X).$$

Here, $X_{P_0}$ describes the polytope of the relaxed version of the original formulation (this corresponds to the LP formulation for the QSAP) and $conv(X)$ is the convex hull of the set of the feasible solutions $X$ of the original QSAP. The other formulations $X_{P_i}$ characterize stepwise improving polyhedral formulations that are gained by the RLT steps.

The theory of the RLT has made quite some progress in the last years, making it one of the most successful approaches for generating lower bounds for QAPs.

## 5.2 Literature Survey

A general overview of the RLT method and the theory that it is based on is presented in [56] and [57]. The theory of the RLT made promising progress for the QAP in the last years. The application of the RTL to the QAP and related problems is studied in detail in [65]. Special techniques, like the RLT-2 formulation ([22]) and the RLT-3 formulation ([24]) provide some of the best bounds known for hard QAP test instances. A further RLT-2 lower bound approach is presented in [1].

For the Generalized Quadratic Assignment Problem (GQAP), which is a generalized form of both the QAP and the QSAP, solution strategies based on RLT approaches are presented in [23] and [43].

A different approach that can be combined with the RLT is the reduction approach of Billionet et al. from [5]. Here, the authors present a strategy to change the objective function by extracting a constant term that is a lower bound of the QSAP. This work is based on the results of [8] and [25] that deal with the theory of pseudo-Boolean optimization.

Furthermore, a theoretic result for the RLT on QSAPs is presented in [54]. This work arises from the results of this thesis and it contains a shortened version of Section 5.6. Here, we construct a problem structure that is responsible for untight solutions and we prove its minimality.

## 5.3   RLT for Quadratic Assignment Problems

In Section 3.8, we presented some promising lower bound results of the RLT for the QAP. Adams et al. [1] and Hahn et al. [24] used Lagrangian Relaxation techniques and Dual Ascent strategies to generate lower bounds for the QAP that are currently state-of-the-art. These bounds are the best ones known for some of the main instances from QAPLib (cf. Section 3.8 and [10]).

The QAP describes the task to assign $m$ objects to $m$ locations, where only one object may be located at each location and a quadratic objective function should be minimized. The Integer Program (IP) formulation can be described with the index set $M = \{1, ..., m\}$ as follows:

$$(\text{IP}_{\text{QAP}}) \ \min \sum_{i,j \in M} b_{ij} x_{ij} + \sum_{\substack{i,k \in M \\ i < k}} \sum_{\substack{j,l \in M \\ j \neq l}} c_{ijkl} x_{ij} x_{kl}$$

$$s.t. \sum_{i \in M} x_{ij} = 1 \quad \forall j \in M,$$

$$\sum_{j \in M} x_{ij} = 1 \quad \forall i \in M,$$

$$x_{ij} \in \{0,1\} \quad \forall i, j \in M.$$

The RLT replaces the non-linear quadratic terms in the objective function by introducing new variables. To keep the structure of the problem, additional constraints for these new variables are introduced. Furthermore, we relax the integer condition of the $x_{ij}$ variables. Thus, we get the new formulation

$$(\text{RLT-1}_{\text{QAP}}) \ \min \sum_{i,j \in M} b_{ij} x_{ij} + \sum_{\substack{i,k \in M \\ i < k}} \sum_{\substack{j,l \in M \\ j \neq l}} c_{ijkl} y_{ijkl}$$

$$s.t. \sum_{i \in M} x_{ij} = 1 \quad \forall j \in M,$$

$$\sum_{j \in M} x_{ij} = 1 \quad \forall i \in M,$$

$$\sum_{\substack{i \in M \\ i \neq k}} y_{ijkl} = x_{kl} \quad \forall j, k, l \in M \ (j \neq l),$$

$$\sum_{\substack{j \in M \\ j \neq l}} y_{ijkl} = x_{kl} \quad \forall i, k, l \in M \ (i \neq k),$$

$$y_{ijkl} = y_{klij} \quad \forall i, j, k, l \in M \ (i < k, \ j \neq l),$$

$$x_{ij} \geq 0 \quad \forall i, j \in M,$$

$$y_{ijkl} \geq 0 \quad \forall i, j, k, l \in M \ (i \neq k, \ j \neq l).$$

This formulation provides lower bounds for the QAP, since the integer condition of the solutions is relaxed. Note that every solution of the IP formulation is also feasible for the RLT formulation.

For higher RLT levels, new variables are introduced (e.g., for the RLT-2 formulation we need new $z_{ijklpq}$ variables). The highest RLT level that was to the best of our knowledge used so far is the level 3 RLT formulation by Hahn et al. in [24].

## 5.4 RLT for QSAPs

In the previous section, the RLT for the QAP was introduced. Most of the basic ideas can also be applied to the QSAP. The QSAP has fewer constraints and as a consequence also a larger search space compared to the QSAP.

In most theoretic analyses of the QSAP, a standard linearization technique is presented. This approach corresponds to the first level of the RLT and is presented in the following subsection.

Note that we focus our research on the RLT for the symmetric QSAP. But all ideas of this chapter can also be applied to the asymmetric QSAP by slightly changing the formulations.

## 5.4.1 The RLT-1 Formulation

The RLT-1 approach that linearizes a QSAP is known as the standard linear program relaxation of the QSAP. In this thesis, it was presented in Section 3.4.2. The idea behind this RLT-1 formulation is as follows:

Given the sets $M = \{1, ..., m\}$ and $N_i = \{1, ..., n_i\}$ for all $i \in M$ and the IP-formulation of the QSAP

$$(\text{IP1}) \ \min \sum_{i \in M} \sum_{j \in N_i} b_{ij} x_{ij} + \sum_{\substack{i,k \in M \\ i<k}} \sum_{\substack{j \in N_i \\ l \in N_k}} c_{ijkl} x_{ij} x_{kl}$$

$$s.t. \sum_{j \in N_i} x_{ij} = 1 \quad \forall i \in M,$$

$$x_{ij} \in \{0,1\} \quad \forall i \in M, \ \forall j \in N_i,$$

the RLT-1 formulation arises from linearizing the products $x_{ij} \cdot x_{kl}$ with new variables $y_{ijkl} := x_{ij} \cdot x_{kl}$ and by adding new constraints. These constraints are generated by multiplying the given constraints

$$\sum_{j \in N_i} x_{ij} = 1$$

by $x_{kl}$ (for all $k \in M \setminus \{i\}$ and for all $l \in N_k$). This results in the new constraints

$$\sum_{j \in N_i} y_{ijkl} = x_{kl} \quad \forall i, k \in M \ (i < k), \ \forall l \in N_k \text{ and}$$

$$\sum_{l \in N_k} y_{ijkl} = x_{ij} \quad \forall i, k \in M \ (i < k), \ \forall j \in N_i.$$

By relaxing the integer condition of the $x_{ij}$ variables, we get the linearized formulation

$$(\text{RLT-1}) \ \min \sum_{i=1}^{m} \sum_{j=1}^{n_i} b_{ij} x_{ij} + \sum_{i=1}^{m-1} \sum_{k=i+1}^{m} \sum_{j=1}^{n_i} \sum_{l=1}^{n_k} c_{ijkl} y_{ijkl}$$

$$s.t. \sum_{j=1}^{n_i} x_{ij} = 1 \quad \forall i \in M,$$

$$\sum_{j=1}^{n_i} y_{ijkl} = x_{kl} \quad \forall i, k \in M \ (i < k), \ \forall l \in N_k,$$

$$\sum_{l=1}^{n_k} y_{ijkl} = x_{ij} \quad \forall i, k \in M \ (i < k), \ \forall j \in N_i,$$

$$x_{ij} \geq 0 \quad \forall i \in M, \ \forall j \in N_i,$$

$$y_{ijkl} \geq 0 \quad \forall i, k \in M \ (i < k), \ \forall j \in N_i, \ \forall l \in N_k.$$

In this RLT-1 formulation, we have

$$\sum_{i=1}^{m} n_i \quad \text{new variables and}$$

$$\sum_{i=1}^{m-1} \sum_{k=i+1}^{m} n_i \cdot n_k \quad \text{new sum constraints.}$$

Note that in the asymmetric case, which includes both variables, $y_{ijkl}$ and $y_{klij}$, we replace the constraints

$$\sum_{j=1}^{n_i} y_{ijkl} = x_{kl} \quad \forall i, k \in M \ (i < k), \ \forall l \in N_k,$$

$$\sum_{l=1}^{n_k} y_{ijkl} = x_{ij} \quad \forall i, k \in M \ (i < k), \ \forall j \in N_i$$

by

$$\sum_{j=1}^{n_i} y_{ijkl} = x_{kl} \quad \forall i, k \in M \ (i \neq k), \ \forall l \in N_k,$$

$$y_{ijkl} = y_{klij} \quad \forall i, k \in M \ (i \neq k), \ \forall j \in N_i, \ \forall l \in N_k.$$

The lower bounds for the QSAP that are calculated with the RLT-1 formulation are promising for small test instances, but the gaps for bigger problem sizes are unfortunately too large for practical purposes. Thus, better relaxations are desirable. The results of the RLT-1 formulation for various test instances are presented in Section 6.3.2.

## 5.4.2 The RLT-2 Formulation

For a sharper polyhedral formulation of the problem, the level 2 RLT adds new variables $z_{ijklpq}$ into the model. These variables arise from multiplying the given constraints

$$\sum_{j=1}^{n_i} y_{ijkl} = x_{kl} \quad \text{and} \quad \sum_{l=1}^{n_k} y_{ijkl} = x_{ij}$$

by $x_{pq}$. Furthermore, we replace the product

$$y_{ijkl} \cdot x_{pq} = x_{ij} \cdot x_{kl} \cdot x_{pq} \quad (i < k)$$

by

$$\begin{aligned} z_{ijklpq}, &\text{ if } i < k < p, \\ z_{ijpqkl}, &\text{ if } i < p < k \text{ or} \\ z_{pqijkl}, &\text{ if } p < i < k. \end{aligned}$$

The new formulation is as follows:

(RLT-2)

$$\min \sum_{i=1}^{m} \sum_{j=1}^{n_i} b_{ij} x_{ij} + \sum_{i=1}^{m-1} \sum_{k=i+1}^{m} \sum_{j=1}^{n_i} \sum_{l=1}^{n_k} c_{ijkl} y_{ijkl}$$

$$s.t. \sum_{j=1}^{n_i} x_{ij} = 1 \quad \forall i \in M,$$

$$\sum_{j=1}^{n_i} y_{ijkl} = x_{kl} \quad \forall i,k \in M \ (i < k), \ l \in N_k,$$

$$\sum_{l=1}^{n_k} y_{ijkl} = x_{ij} \quad \forall i,k \in M \ (i < k), \ j \in N_i,$$

$$\sum_{j=1}^{n_i} z_{ijklpq} = y_{klpq} \quad \forall i,k,p \in M \ (i < k < p), \ \forall l \in N_k, \ \forall q \in N_p,$$

$$\sum_{l=1}^{n_k} z_{ijklpq} = y_{ijpq} \quad \forall i,k,p \in M \ (i < k < p), \ \forall j \in N_i, \ \forall q \in N_p,$$

$$\sum_{q=1}^{n_p} z_{ijklpq} = y_{ijkl} \quad \forall i,k,p \in M \ (i < k < p), \ \forall j \in N_i, \ \forall l \in N_k,$$

$$x_{ij} \geq 0 \ \forall i \in M, \ \forall j \in N_i,$$

$$y_{ijkl} \geq 0 \ \forall i,k \in M \ (i < k), \ \forall j \in N_i, \forall l \in N_k,$$

$$z_{ijklpq} \geq 0 \ \forall i,k,p \in M \ (i < k < p), \ \forall j \in N_i, \forall l \in N_k, \forall q \in N_p.$$

The lower bounds generated by the RLT-2 formulation cannot be worse than the ones generated by the RLT-1 formulation, since all feasible solutions of the RLT-2 formulation are also feasible for the RLT-1 formulation.

We present detailed results of the RLT-2 formulation in Section 6.3.3. The solution quality of the formulation is promising, but it is achieved for the price of high computational efforts. Solving the RLT-2 formulation with an LP solver takes in most cases even longer than the time that the MIP solver needs to solve the exact MIP formulation (we solve both formulations with Ilog Cplex 11.2). This is caused by the high number of new variables and constraints in the RLT-2 formulation. The RLT-2 formulation contains

$$\sum_{\substack{i,k,p\in M \\ i<k<p}} n_i \cdot n_k \cdot n_p \quad \text{new variables and}$$

$$\sum_{\substack{i,k,p\in M \\ i<k<p}} n_i \cdot n_k + n_i \cdot n_p + n_k \cdot n_p \quad \text{new sum constraints}$$

in comparison to the RLT-1 formulation.

For the asymmetric RLT-2 formulation, we add the constraints

$$\sum_{j=1}^{n_i} z_{ijklpq} = y_{klpq} \quad \forall i,k,p \in M \ (i \neq k, i \neq p, k \neq p),$$

$$\forall l \in N_k, \ \forall q \in N_p,$$

and

$$z_{ijklpq} = z_{ijpqkl} = z_{klijpq} = z_{klpqij} = z_{pqijkl} = z_{pqklij}$$
$$\forall i,k,p \in M \ (i \neq k, i \neq p, k \neq p, i < k < p),$$
$$\forall j \in N_i, \ \forall l \in N_k, \ \forall q \in N_p.$$

to the asymmetric RLT-1 formulation

## 5.4.3 Level-$t$ RLT

As described in the introduction of this chapter, the RLT constructs an $n$ level hierarchy. Up to now, we introduced the RLT formulations for the first two levels. In this subsection, a new general notation for the level $t$ formulation is presented. This notation is introduced in the following definition.

**Definition 5.1 (RLT variable)** *We denote the RLT variables that are introduced during the level $t$ linearization step of the RLT for the QSAP by*

$$\vartheta^{(t)}_{i_1 j_1, \ldots, i_{t+1} j_{t+1}} \quad with$$

$$i_1, \ldots, i_{t+1} \in M \ (i_1 < \ldots < i_{t+1}), \ j_s \in N_{i_s} \ \forall s \in \{1, \ldots, t+1\}.$$

*Each RLT-t variable has $2 \cdot (t+1)$ indices. For a consistent notation, we define for $t = -1$*

$$\vartheta^{(-1)} = 1.$$

*We denote the combined set of all RLT variables $\vartheta^{(0)}, \ldots, \vartheta^{(t)}$ by $\Theta^t$.*

For the already introduced RLT-1 and RLT-2 formulations, the new $\vartheta^{(0)}$, $\vartheta^{(1)}$ and $\vartheta^{(2)}$ variables correspond to the $x_{ij}$, $y_{ijkl}$ and $z_{ijklpq}$ variables in the following way:

$$x_{ij} = \vartheta^{(0)}_{ij}, \quad y_{ijkl} = \vartheta^{(1)}_{ij,kl}, \quad z_{ijklpq} = \vartheta^{(2)}_{ij,kl,pq}.$$

In each reformulation step of the RLT, new constraints are added to the formulation. We introduce these constraints for the general RLT-$t$ formulation in the following definition.

**Definition 5.2 (RLT-$t$ constraints, RLT-$t^{\leq}$ constraints)** *We call the constraints that are introduced in the level $t$ reformulation step RLT-$t$ constraints. Such an RLT-$t$ constraint has the form*

$$\sum_{j_k=1}^{n_k} \vartheta^{(t)}_{i_1 j_1, \dots, i_{t+1} j_{t+1}} = \vartheta^{(t-1)}_{i_1 j_1, \dots, i_{k-1} j_{k-1}, i_{k+1} j_{k+1}, \dots, i_{t+1} j_{t+1}}$$

$$k \in \{1, \dots, t+1\}, \ i_1, \dots, i_{t+1} \in M \ (i_1 < \dots < i_{t+1}),$$

$$j_s \in N_{i_s} \ (s \in \{1, \dots, t+1\} \setminus \{k\}).$$

*For an easier notation, we denote the set of all RLT-$\tau$ constraints $(0 \leq \tau \leq t)$ as RLT-$t^{\leq}$ constraints.*

We simplify the naming of feasible solutions in the following definition.

**Definition 5.3 (RLT-$t$ feasible solution, RLT-$t$ solution)** *We call an assignment $\sigma : \Theta^t \to [0,1]$ of all $\vartheta^{(0)}$, ..., $\vartheta^{(t)}$ variables an RLT-$t$ feasible solution of an RLT-$t$ formulation, if the variables fulfill all RLT-$t^{\leq}$ constraints. As a shorter notation, we name such a solution an RLT-$t$ solution.*

The goal of an RLT-$t$ formulation is to find an optimal assignment $\sigma$ to all $\vartheta$-variables so that all RLT-$t^{\leq}$ constraints are satisfied. The complete level $t$

RLT formulation is as follows:

$$(\text{RLT-t}) \quad \min \sum_{i=1}^{m} \sum_{j=1}^{n_i} b_{ij} x_{ij} + \sum_{i=1}^{m-1} \sum_{k=i+1}^{m} \sum_{j=1}^{n_i} \sum_{l=1}^{n_k} c_{ijkl} y_{ijkl}$$

$$s.t. \sum_{j_k=1}^{n_k} \vartheta_{i_1 j_1, \dots, i_k j_k, \dots, i_{\tau+1} j_{\tau+1}}^{(\tau)} = \vartheta_{i_1 j_1, \dots, i_{k-1} j_{k-1}, i_{k+1} j_{k+1}, \dots, i_{\tau+1} j_{\tau+1}}^{(\tau-1)}$$

$$\forall \tau \in \{0, \dots, t\}, \ \forall k \in \{1, \dots, \tau+1\},$$

$$\forall i_1, \dots, i_{\tau+1} \in M \ (i_1 < \dots < i_{\tau+1}),$$

$$\forall j_s \in N_s \ (s \in \{1, \dots, \tau+1\} \setminus \{k\}),$$

$$\vartheta_{i_1 j_1, \dots, i_{\tau+1} j_{\tau+1}}^{(\tau)} \geq 0$$

$$\forall \tau \in \{0, \dots, t\},$$

$$\forall i_1, \dots, i_{\tau+1} \in M \ (i_1 < \dots < i_{\tau+1}),$$

$$\forall j_s \in N_s \ (s \in \{1, \dots, \tau+1\}).$$

### 5.4.4 Including Constraints in the RLT Formulation

Similar to the handling of constraints for the MIP and the LP formulation, we set the variables $\vartheta_{ij}^{(0)}$ or $\vartheta_{ijkl}^{(1)}$ that correspond to forbidden shifts for a line or to a forbidden line shift pair combination to zero. Thus, all solutions of this constrained RLT formulation are also lower bounds for the constrained QSAP.

## 5.5 Interpretation of the RLT on Graphs

The following notation is based on the definition of the QSAP-graph (Definition 3.7) from Chapter 3. For a representation of an RLT formulation on graphs, each $x_{ij}$ variable corresponds to a vertex $v_{ij}$. An edge between two vertices $v_{ij}$ and $v_{kl}$ corresponds to the linearized variable $y_{ijkl}$. The graph that describes a QSAP is defined by

$$G_{QSAP} = (V_{QSAP}, E_{QSAP}) \text{ with}$$
$$V_{QSAP} = \{v_{ij} \mid i \in M, \ j \in N_i\},$$
$$E_{QSAP} = \{e_{ijkl} = \{v_{ij}, v_{kl}\} \mid i, k \in M \ (i < k), \ j \in N_i, \ l \in N_k\}.$$

For a better visualization, we regard rectangular visualizations of the QSAP graphs. Here, each element of the set $M$ corresponds to a row and the set $N_i$ corresponds to the vertices of row $i$. Figure 5.1 shows such a rectangular graph representation of a QSAP.

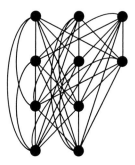

Figure 5.1: Rectangular representation of a graph $G_{QSAP}$.

In addition, we need the solution graphs from Definition 3.8. For a feasible solution $\sigma$ of an RLT formulation, such a solution graph has the form

$$G_{sol} = (V_{sol}, E_{sol}) \text{ with}$$
$$V_{sol} = \{v_{ij} \in V_{QSAP} \mid \vartheta_{ij}^{(0)} = x_{ij} > 0\},$$
$$E_{sol} = \{e_{ijkl} \in E_{QSAP} \mid \vartheta_{ij,kl}^{(1)} = y_{ijkl} > 0\}.$$

For an interpretation of the general $\vartheta^{(t)}$-variables with $t \geq 2$, we have to introduce the concept of hypergraphs.

**Definition 5.4 (Hypergraph, Hyperedge)** *A hypergraph is a generalization of a graph, where edges can also connect more than two vertices. Thus, a hypergraph is a pair $H = (V, E)$, where $V$ is a set of vertices and $E$ is a set of non-empty subsets of $V$. These subsets $e \subseteq V$ are called hyperedges.*

If we regard the QSAP-graph as a hypergraph, a $z_{ijklpq}$ variable corresponds to a hyperedge between the vertices $v_{ij}$, $v_{kl}$ and $v_{pq}$. Accordingly, a $\vartheta^{(t)}$ variable corresponds to a hyperedge with $(t+1)$ vertices.

In this thesis, we regard quadratic problems. This means that the objective function is related to the edges of the graph and not to higher hyperedges. Thus, we will focus our research on ordinary graphs. But even here, these higher $\vartheta^{(t)}$ ($t \geq 2$) variables have a meaning. This is shown in the following lemma.

**Lemma 5.5** *Given is a solution $\sigma$ of an RLT-t formulation. If an RLT variable has a value greater than zero, then the vertices of the corresponding hyperedge must form a clique in the solution graph. Conversely, if a set of vertices forms no t-clique in a solution graph, then the $\vartheta^{(t-1)}$ variable that corresponds to the hyperedge between these vertices must have the assigned value zero.*

**Proof:**

Let $\vartheta^{(\tau)}_{i_1 j_1, \ldots, i_{\tau+1} j_{\tau+1}}$ ($1 \leq \tau \leq t$) be an RLT variable with an assigned value greater than zero in the given solution $\sigma$. Since all RLT-$t^{\leq}$ constraints must be satisfied, it follows directly that the inequalities

$$\vartheta^{(1)}_{i_u j_u, i_v j_v} > 0 \quad \forall u, v \in \{1, \ldots, \tau+1\} \ (u < v) \ \text{ and}$$
$$\vartheta^{(0)}_{i_u j_u} > 0 \quad \forall u \in \{1, \ldots, \tau+1\}$$

hold. For the edges that correspond to these $\vartheta^{(1)}_{i_u j_u, i_v j_v}$ variables in the solution graph $G_{sol}$, we have

$$e_{i_u j_u i_v j_v} \in E_{sol},$$

because of $\vartheta^{(1)}_{i_u j_u, i_v j_v} > 0$. Therefore, the vertices $v_{ij}$ that correspond to the $\vartheta^{(0)}_{i_u j_u}$ variables ($u \in \{1, \ldots, \tau+1\}$) form a clique in the solution graph $G_{sol}$.

Conversely, it follows via contraposition that if $\tau$ vertices form no $\tau$-clique, then the corresponding RLT-$(\tau-1)$ variable must have the assigned value zero.

$\square$

Note that the reversed statement is not necessarily true. This means that there can be a $(\tau+1)$-clique in a solution graph for which the corresponding RLT-$\tau$ variable has value zero.

The meaning of the RLT-$t$ constraints for the graph formulation are as follows. Given a solution graph of a feasible RLT-2 solution, the constraints ensure that for each edge $e_{ijkl}$ with $y_{ijkl} > 0$ and for every $p \in M \setminus \{i, k\}$ there is at least one vertex triple $\{v_{ij}, v_{kl}, v_{pq}\}$ for which the vertices are pairwise adjacent. These vertices form a 3-clique in $G_{sol}$ and we have

$$z_{ijklpq} > 0 \text{ if } (i < k < p),$$
$$z_{ijpqkl} > 0 \text{ if } (i < p < k) \text{ or}$$
$$z_{pqijkl} > 0 \text{ if } (p < i < k).$$

This result can be generalized for all RLT levels.

**Lemma 5.6** *Given are a feasible solution $\sigma$ of an RLT-t formulation and a corresponding solution graph $G_{sol}$. For every $\tau$ with $1 \leq \tau \leq t$ and for each set $\{i_1, ..., i_{\tau+1}\} \in M$ with $i_1 < ... < i_{\tau+1}$, we regard the induced subgraph that consists only of the rows with the indices $i_1, ..., i_{\tau+1}$. This subgraph consists of a union of $(\tau + 1)$-cliques.*

**Proof:**

Given is a set of row indices $\{i_1, ..., i_{\tau+1}\}$ with $i_1 < ... < i_{\tau+1}$ ($\tau < t$) and a feasible solution $\sigma$ that fulfills all RLT-$t^{\leq}$ constraints. From the RLT-0 constraints we know that

$$\sum_{j_s=1}^{n_{i_s}} \vartheta_{i_s j_s}^{(0)} = 1 \quad \forall s \in \{1, ..., t+1\}$$

holds. Thus, there must be at least one RLT-0 variable $x_{ij}$ with $x_{ij} > 0$ for every $i \in M$. We can extend this RLT-0 constraint with the RLT-1 constraints of the form

$$\sum_{j_u=1}^{n_{i_u}} \vartheta_{i_s j_s, i_u j_u}^{(1)} = \vartheta_{i_s j_s}^{(0)}$$

into

$$\sum_{j_s=1}^{n_{i_s}} \vartheta_{i_s j_s}^{(0)} = \sum_{j_s=1}^{n_{i_s}} \sum_{j_u=1}^{n_{i_u}} \vartheta_{i_s j_s, i_u j_u}^{(1)} = 1 \quad \forall s, u \in \{1, ..., t+1\} \ (s < u).$$

By considering the higher RLT constraints, we finally get the equation

$$\sum_{j_1=1}^{n_{i_1}} \cdots \sum_{j_{\tau+1}=1}^{n_{i_{\tau+1}}} \vartheta_{i_1 j_1, ..., i_{\tau+1} j_{\tau+1}}^{(\tau)} = 1.$$

Each of the $\vartheta^{(\tau)}$ variables with $\vartheta_{i_1 j_1, ..., i_{\tau+1} j_{\tau+1}}^{(\tau)} > 0$ induces a clique in the solution graph of $\sigma$. The corresponding $\vartheta^{(0)}$ and $\vartheta^{(1)}$ variables sum up to one. Thus, no further vertices or edges than the ones that are induced by the $\vartheta^{(\tau)}$ variables may exist in $G_{sol}$. This proves the lemma.

□

For the connection between the solutions of an RLT formulation and the corresponding graph formulation, we introduce the following definition.

**Definition 5.7** *We say that a graph $G = (V, E)$ that is a subgraph of $G_{QSAP}$ satisfies the RLT-$t^\leq$ constraints, if there exists a variable assignment $\sigma$ to all RLT variables that satisfies the RLT-$t^\leq$ constraints and for which*

$$\sigma(\vartheta_{ij,kl}^{(1)}) > 0$$

*holds if and only if $e_{ijkl} \in E$.*

Note that all RLT variables whose corresponding vertices do not form a clique in $G_{sol}$ must have the assigned value zero in such a feasible variable assignment.

We will use this definition to generate feasible RLT solutions that do not match to a solution of a QSAP. Here, we search for graphs that satisfy the RLT-$t^\leq$ constraints and that do not consist of a union of $m$-cliques.

# 5.6 Finding Tight RLT Formulations

While comparing the results of the RLT-1 and the RLT-2 formulation, it is noticeable that the optimal solution value of the RLT-2 formulation is often the optimal solution value of the QSAP, while the solution of the RLT-1 formulation has large gaps. Therefore, an analysis of the influence of the RLT level on the structure of the solution graphs stands to reason. Additionally, it is interesting to find those QSAPs for which the RLT-2 formulation does not generate an optimal solution. This may help to find strategies that avoid big gaps while generating lower bounds.

We start by analyzing optimal solutions of RLT-1 formulations that do not have the same values as the optimal QSAP solution. Here, the solution graphs often contain a typical structure that is shown in Figure 5.2. This graph contains no 3-clique but it satisfies all RLT-$1^\leq$ constraints.

A feasible variable assignment for the vertices $v_{ij}$ and the edges $e_{ijkl}$ of this graph is $\vartheta_{ij}^{(0)} = \vartheta_{ijkl}^{(1)} = \frac{1}{2}$. We assign to all other RLT variables of the

Figure 5.2: Example of a graph that satisfies the RLT-$1^{\leq}$ constraints but that does not correspond to a solution of a QSAP, since it contains no 3-clique.

corresponding QSAP with $|M| = 3$ and $|N_1| = |N_2| = |N_3| = 2$ the value zero. This variable assignment fulfills all RLT-$1^{\leq}$ constraints

Such a graph structure that satisfies the RLT-2 constraints but that does not correspond to a QSAP solution is not easy to find. But we construct in [54] a feasible graph in terms of the RLT-$2^{\leq}$ constraints for a problem of size $|M| = 4$ and $|N_i| = 3$ for all $i \in M$ that contains no 4-clique and is therefore not a feasible solution of the original problem. This graph is presented in Figure 5.3.

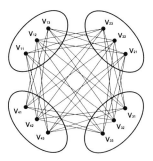

Figure 5.3: Example of a graph that satisfies the RLT-$2^{\leq}$ constraints but that is not a solution of a QSAP. For a better visualization, the graph is not presented in the rectangular form.

A feasible variable assignment for the corresponding $\vartheta^{(t)}$ variables ($t \in \{0, ..., 2\}$) that satisfies the RLT-$2^{\leq}$ constraints is $\vartheta^{(0)}_{ij} = \frac{1}{3}$ and $\vartheta^{(1)}_{ijkl} = \vartheta^{(2)}_{ijklpq} = \frac{1}{6}$ for the vertices, edges and 3-cliques of this graph. All other RLT variables get the assigned value zero.

This graph can be constructed by removing three disjoint 4-cliques from the complete multipartite QSAP-graph $G_{QSAP}$ with $|M| = 4$ and $|N_i| = 3$ for all $i \in M$ that was shown in Figure 3.3. The removed cliques are presented in Figure 5.4.

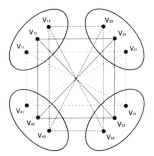

Figure 5.4: Three 4-cliques (marked by different dashed lines) that are removed from $G_{QSAP}$ to generate the graph from Figure 5.3.

It can be seen for the graph from Figure 5.2 that it is also constructed from a QSAP-graph by a removal of $m$-cliques ($m = 3$). This is shown in Figure 5.5.

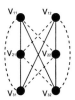

Figure 5.5: Graph from Figure 5.2 with missing edges as dashed lines. The dashed lines form two disjoint 3-cliques.

From these graphs the question arises, whether there exists a general graph structure that satisfies all RLT-$t^{\leq}$ constraints for a certain level $t$ but that contains no $m$-clique ($|M| = m$ denotes the size of the QSAP).

The goal of such a consideration is to find the RLT level $t$ for a given problem size so that the RLT-$t$ formulation returns the optimal solution value as a lower bound for the corresponding QSAP. This is done by preventing a possible occurrence of a minimal form of these graph structures.

## 5.6.1 Concept of Tight RLT Formulations

The RLT formulations are not only capable of generating good lower bounds, they can also generate the optimal solution value of the QSAP if the RLT level is large enough. We call an RLT solution tight, if it has no gap to the optimal QSAP solution value.

**Definition 5.8 (Tight RLT Solution)** *For a given QSAP instance (represented by the sets $M$ and $N_i$ for all $i \in M$ and an objective function $f$) and an RLT level $t$, an RLT-t feasible solution $\sigma$ is called a tight RLT solution, if the objective function value $f(\sigma)$ is the same as the optimal solution value of the QSAP.*

Note that an optimal solution of an RLT formulation that has the same value as the optimal solution of the underlying QSAP must not have the same variable assignment. An easy example for such a consideration can be given, if the QSAP has two solutions with the same optimal value. Every linear combination of these two solutions is also a feasible solution of the RLT-$t$ formulation, independently of the RLT level $t$. But since we are primarily interested in good lower bounds and not in the solution structure, such a linear combination does not interfere with our goals. In addition, it is easy to construct the individual solutions that the linear combination consists of.

The information, whether an RLT solution of a certain QSAP is tight, is clearly useful. But in this section, we are interested in the quality of an RLT-$t$ formulation in general and not only in the solution quality of the optimal RLT solution for a single QSAP instance. A concept for such a quality measure is presented in the following definition.

**Definition 5.9 (Tight RLT Formulation)** *For a given QSAP size (represented by the set $M$ and the sets $N_i$ for all $i \in M$), an RLT-t formulation is called tight, if the optimal solution of the RLT-t formulation is tight for every QSAP of the given size, independently of the coefficients $b_{ij}$ and $c_{ijkl}$ in the objective function.*

We are interested in the smallest RLT level $t$ for a given QSAP size so that the RLT-$t$ formulation is tight. For a further analysis of the problem of finding this smallest RLT level, we need the following lemma.

**Lemma 5.10** *Each linear combination of RLT-t solutions is also an RLT-t solution.*

**Proof:**

Let $\mu = \{\mu^{(0)}, ..., \mu^{(t)}\}$ and $\nu = \{\nu^{(0)}, ..., \nu^{(t)}\}$ be the variable assignments of two feasible solutions of an RLT-$t$ formulation, meaning that the RLT variables $\mu^{(\tau)}$ and $\nu^{(\tau)}$ $(0 \leq \tau \leq t)$ satisfy all RLT-$t^{\leq}$ constraints. We show that $\vartheta = (a_1 \cdot \mu + a_2 \cdot \nu)$ with $(a_1 + a_2) = 1$ also satisfies all RLT-$t^{\leq}$ constraints. The new assignment $\vartheta$ is defined by

$$\vartheta^{(\tau)}_{i_1 j_1, ..., i_{\tau+1} j_{\tau+1}} = a_1 \cdot \mu^{(\tau)}_{i_1 j_1, ..., i_{\tau+1} j_{\tau+1}} + a_2 \cdot \nu^{(\tau)}_{i_1 j_1, ..., i_{\tau+1} j_{\tau+1}}$$

for all $0 \leq \tau \leq t$, for all $i_s \in M$ $(i_u < i_v$ if $u < v)$ and for all $j_s \in N_{i_s}$ $(s \in \{1, ..., \tau+1\})$. Given an RLT-$\tau$ constraint $(1 \leq \tau \leq t)$, we have the two equations

$$\sum_{j_k=1}^{n_k} \mu^{(\tau)}_{i_1 j_1, ..., i_k j_k, ..., i_{\tau+1} j_{\tau+1}} = \mu^{(\tau-1)}_{i_1 j_1, ..., i_{k-1} j_{k-1}, i_{k+1} j_{k+1}, ..., i_{\tau+1} j_{\tau+1}},$$

$$\sum_{j_k=1}^{n_k} \nu^{(\tau)}_{i_1 j_1, ..., i_k j_k, ..., i_{\tau+1} j_{\tau+1}} = \nu^{(\tau-1)}_{i_1 j_1, ..., i_{k-1} j_{k-1}, i_{k+1} j_{k+1}, ..., i_{\tau+1} j_{\tau+1}}.$$

By adding these two equations with factors $a_1$ and $a_2$, we get

$$\sum_{j_k=1}^{n_k} (\vartheta)^{(\tau)}_{i_1 j_1, ..., i_k j_k, ..., i_{\tau+1} j_{\tau+1}} = (\vartheta)^{(\tau-1)}_{i_1 j_1, ..., i_{k-1} j_{k-1}, i_{k+1} j_{k+1}, ..., i_{\tau+1} j_{\tau+1}}.$$

Thus, $\vartheta = (a_1 \cdot \mu + a_2 \cdot \nu)$ satisfies the same RLT-$\tau$ constraint.

The RLT-0 constraint plays a special role here, since it normalizes the variables of a solution. For this constraint, we get the equations

$$\sum_{j_1=1}^{n_1} \mu^{(0)}_{i_1 j_1} = 1,$$

$$\sum_{j_1=1}^{n_1} \nu^{(0)}_{i_1 j_1} = 1.$$

By adding these two equations with the factors $a_1$ and $a_2$, we get

$$\sum_{j_k=1}^{n_k} (\vartheta)^{(0)}_{i_1 j_1} = a_1 + a_2 = 1.$$

Therefore, the variable assignment $\vartheta$ satisfies all RLT-$t^{\leq}$ constraints.

We can construct any linear combination of solutions by stepwise building weighted sums of two solutions. This proves the lemma.

$\square$

In Figure 5.2 and Figure 5.3, we presented two graphs of RLT solutions that do not correspond to solutions or linear combinations of solutions of a QSAP. With these graphs it is possible to create QSAPs of the given sizes ($|M| = 3$, $|N_i| = 2$ for all $i \in M$ or $|M| = 4$, $|N_i| = 3$ for all $i \in M$) for which the RLT-1 formulation or the RLT-2 formulation return untight lower bounds.

An example of such a QSAP can be given for both sizes by the following objective function. We set all coefficients $b_{ij}$ to zero and all coefficients $c_{ijkl}$ to one except for those coefficients that correspond to the edges of the graphs of Figure 5.2 or Figure 5.3. The coefficients of these edges also get the value zero. With these objective functions, we get two QSAPs whose optimal RLT-1 or RLT-2 solutions have value zero. But the optimal QSAP solutions have both value one. Thus, the RLT solutions are not tight and it follows that the RLT-1 or RLT-2 formulations are untight for the given QSAP sizes.

A general form of this graph structure that generates untight formulations can be found for any RLT level. The following definition introduces this graph structure.

**Definition 5.11** ($G_{RLT}^t$)  *For the graph $G_{QSAP} = (V_{QSAP}, E_{QSAP})$ with $M = \{1, ..., t+2\}$ and $N_i = \{1, ..., t+1\}$ for all $i \in M$, the graph $G_{RLT}^t$ is defined by*

$$G_{RLT}^t = \left(V_{RLT}^t, E_{RLT}^t\right) \quad with$$
$$V_{RLT}^t = V_{QSAP} = \{v_{ij} \mid i \in M, j \in N_i\},$$
$$E_{RLT}^t = \left\{ e_{ijkl} \in E_{QSAP} \mid \begin{array}{l} i, k \in \{1, ..., t+2\} \ (i < k), \\ j, l \in \{1, ..., t+1\} \ (j \neq l) \end{array} \right\}.$$

*In addition, we define the variable assignment $\sigma$ for the graph $G_{RLT}^t$ by*

$$\sigma(\vartheta_{ij}^{(0)}) = \frac{1}{t+1} \quad \forall i \in M, \ \forall j \in N_i$$

*and*

$$
\sigma(\vartheta^{(\tau)}_{i_1 j_1, \dots, i_{\tau+1} j_{\tau+1}}) = 
\begin{cases}
\prod\limits_{s=1}^{\tau+1} \frac{1}{(t+2-s)} & , \text{ if } e_{i_k j_k i_l j_l} \in E^t_{RLT} \\
& \forall k, l \in \{1, \dots, \tau+1\} \ (k < l), \\
\\
0 & , \text{ else}
\end{cases}
$$

$\forall \tau \in \{1, \dots, t\},$

$\forall i_1, \dots, i_{\tau+1} \in \{1, \dots, t+2\} \ (i_1 < \dots < i_{\tau+1}),$

$\forall j_1, \dots, j_{\tau+1} \in \{1, \dots, t+1\}.$

In the following lemma, we show that these $G^t_{RLT}$ graphs correspond to solutions of an RLT-$t$ formulation for a QSAP of size $|M| = (t+2)$ and $|N_i| = (t+1)$ for all $i \in M$.

**Lemma 5.12** *The graph $G^t_{RLT}$ satisfies all RLT-$t^{\le}$ constraints for a QSAP of size $|M| = (t+2)$ and $|N_i| = (t+1)$ for all $i \in M$.*

**Proof:**

We have to show that all RLT-$\tau$ constraints

$$
\sum_{j_k=1}^{t+1} \vartheta^{(\tau)}_{i_1 j_1, \dots, i_{\tau+1} j_{\tau+1}} = \vartheta^{(\tau-1)}_{i_1 j_1, \dots, i_{k-1} j_{k-1}, i_{k+1} j_{k+1}, \dots, i_{\tau+1} j_{\tau+1}},
$$

with $\tau \in \{0, \dots, t\}$, $k \in \{1, \dots, \tau+1\}$, $i_1, \dots, i_{\tau+1} \in \{1, \dots, t+2\} \ (i_1 < \dots < i_{\tau+1})$ and $j_1, \dots, j_{\tau+1} \in \{1, \dots, t+1\}$ are fulfilled.

With the given variable assignment from Definition 5.11, we get

$$\sum_{j_k=1}^{t+1} \vartheta_{i_1 j_1,\dots,i_{\tau+1} j_{\tau+1}}^{(\tau)} = \sum_{\substack{j_k=1 \\ j_k \notin \{j_1,\dots,j_{k-1},j_{k+1},\dots,j_{\tau+1}\}}}^{t+1} \vartheta_{i_1 j_1,\dots,i_{\tau+1} j_{\tau+1}}^{(\tau)} + \tau \cdot 0$$

$$= ((t+1)-\tau) \cdot \prod_{s=1}^{\tau+1} \frac{1}{(t+2-s)}$$

$$= \frac{(t+1-\tau)}{(t+1)\cdot \dots \cdot (t+2-\tau)\cdot(t+2-(\tau+1))}$$

$$= \frac{1}{t+1} \cdot \dots \cdot \frac{1}{t+2-\tau}$$

$$= \prod_{s=1}^{\tau} \frac{1}{(t+2-s)}$$

$$= \vartheta_{i_1 j_1,\dots,i_{k-1} j_{k-1},i_{k+1} j_{k+1},\dots,i_{\tau+1} j_{\tau+1}}^{(\tau-1)}.$$

This holds for all $\tau \in \{1,\dots,t\}$ and for all $i_1,\dots,i_{t+1} \in \{1,\dots,t+2\}$. For $\tau = 0$, we get

$$\prod_{s=1}^{0} \frac{1}{(t+2-s)} = 1.$$

This proves the lemma.

$\square$

From Lemma 5.12, we know that the graph $G_{RLT}^t$ corresponds to a feasible solution of an RLT-$t$ formulation. But the interesting aspect of these graphs is that they contain no $m$-clique. This is shown in the following lemma.

**Lemma 5.13** *The graph $G_{RLT}^t$ contains no $m$-clique, where $m = (t+2)$ is the number of rows of this graph.*

**Proof:**

Suppose that there exists an $m$-clique in $G_{RLT}^t$ with the vertices

$$C = \{v_{(1)(j_1)},\dots,v_{(t+2)(j_{t+2})}\}.$$

From Definition 5.11, we know that the vertices $v_{(i)(j_i)}$ and $v_{(k)(j_k)}$ $(i \neq k)$ are adjacent in $G_{RLT}^t$ if and only if $j_i \neq j_k$ holds. Thus, $j_1,\dots,j_{t+2}$ must be pairwise different. For the set $I = \{j_1,\dots,j_{t+2}\}$, we have

$$|I| = t+2 > t+1 = |N_i| \quad \forall i \in M.$$

Hence, no $(t + 2)$-clique $C$ can exist in $G^t_{RLT}$, since at least one row with $(t + 2)$ vertices would be necessary to construct such a clique. This proves the lemma.

$\square$

This knowledge about the structure of the $G^t_{RLT}$ graphs can be used to construct QSAPs for which the RLT-$t$ formulation is untight. We already mentioned these examples for the RLT-1 formulation and the RLT-2 formulation. The following lemma introduces a general form of these QSAPs.

**Lemma 5.14** *For every RTL level $t \in \mathbb{N}$, there exists a QSAP instance of size $|M| = (t + 2)$ and $|N_i| = (t + 1)$ for all $i \in M$ for which the graph $G^t_{RLT} = (V^t_{RLT}, E^t_{RLT})$ corresponds to an untight optimal RLT-t solution.*

**Proof:**

To prove the lemma, we construct such a QSAP instance of the given size. The objective function of this QSAP instance is defined by the coefficients

$$b_{ij} = 0 \quad \forall i \in M, \; \forall j \in N_i \quad \text{and}$$

$$c_{ijkl} = \begin{cases} 0, & \text{if } e_{ijkl} \in E^t_{RLT}, \\ 1, & \text{else.} \end{cases}$$

The optimal solution of the RLT-$t$ formulation for this problem instance is zero. In contrast to this, the optimal solution value of the QSAP is one, since at least one edge with weight one must be chosen to construct a $(t+2)$-clique. Thus, the lower bound gained by the RLT-$t$ formulation is not tight.

$\square$

With the help of these $G^t_{RLT}$ graphs, we can determine for each QSAP size the smallest RLT level $t$ that generates a tight RLT-$t$ formulation. But for the proof of the corresponding theorem, we need the following definition.

**Definition 5.15** ($\gamma(m, N)$) *For a QSAP of size $|M| = m$ with the set $N = \{N_1, ..., N_m\}$ ($|N_i| = n_i$ for all $i \in M$), we order the values $n_i$ as a descending sequence $n_{i_1}, ..., n_{i_m}$ so that $n_{i_1} \geq ... \geq n_{i_m}$ ($i_j \neq i_k$ if $j \neq k$) holds. We introduce*

$$\gamma(m, N) = \min\{s \in \mathbb{N} \mid s + 1 > n_{i_{s+2}}\}$$

*as the smallest number $s$ so that no $G^s_{RLT}$ graph can be a subgraph of the corresponding QSAP-graph $G_{QSAP}$.*

In the following theorem, we show that RLT-$\gamma(m, N)$ is the smallest RLT formulation which is tight. Thus, an RLT-$t$ formulation is untight, if $t < \gamma(m, N)$ holds. This result is shown in the following theorem.

**Theorem 5.16 (Smallest Tight RLT Formulation)** *For a given QSAP size $|M| = m$ and $N = \{N_1, ..., N_m\}$, the RLT-$\gamma(m, N)$ formulation is the smallest tight RLT formulation for this problem size.*

## 5.6.2  Proof of the Smallest Tight RLT Formulation Theorem

The proof of Theorem 5.16 consists of several lemmas and it contains two main aspects. The first one is to show that the RLT-$(\gamma(m, N)$-1) formulation is not tight and the second one is to show that the RLT-$\gamma(m, N)$ formulation is tight. Figure 5.6 presents the structure of the proof. Some of the lemmas that are needed were already proven in the previous subsection.

We start by showing that the RLT-$\gamma(m, N)$ formulation is tight. For this, we need a sufficient condition for a formulation to be tight. This condition is presented in the following lemma.

**Lemma 5.17** *Given are a QSAP of size $|M| = m$, $N = \{N_1, ..., N_m\}$ and an RLT-$t$ formulation. If the solution graph $G_{sol}$ that corresponds to an optimal RLT-$t$ solution $\sigma$ consists of a union of $m$-cliques $C_1, ..., C_r$ and if there exist weights $c_1, ..., c_r \in (0, 1]$ for these cliques with $\sum_{s=1}^{r} c_s = 1$ so that the RLT variables of assignment $\sigma$ satisfy*

$$\sigma(\vartheta_{ij}^{(0)}) = \sum_{\substack{s \in \{1,...,r\} \\ v_{ij} \in C_s}} c_s \quad \forall i \in M, \ \forall j \in N_i \ and$$

$$\sigma(\vartheta_{ijkl}^{(1)}) = \sum_{\substack{s \in \{1,...,r\} \\ v_{ij}, v_{kl} \in C_s}} c_s \quad \forall i, k \in M \ (i < k), \ \forall j \in N_i, \ \forall l \in N_k,$$

*then $\sigma$ is a tight solution.*

**Proof:**

Independently of the variable assignment of solution $\sigma$ to the higher RLT variables, we can construct new RLT-$t$ feasible solutions $\sigma'_s$ ($s \in \{1, ..., r\}$)

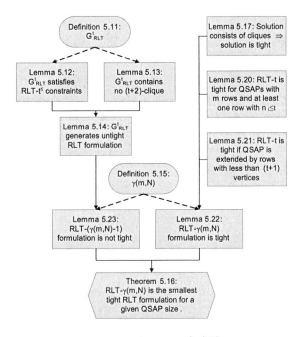

Figure 5.6: Structure of the proof of Theorem 5.16.

with the corresponding variable assignments

$$\sigma'_s\left(\vartheta^{(\tau)}_{i_1j_1,\ldots,i_{\tau+1}j_{\tau+1}}\right) = \begin{cases} 1, & \text{if } v_{i_uj_u} \in C_s \; \forall u \in \{1,\ldots,\tau+1\}, \\ 0, & \text{else} \end{cases} \quad \forall \tau \in \{0,\ldots,t\}.$$

From Lemma 5.10 we know that the linear combination of two RLT-$t$ solutions is also an RLT-$t$ solution. Thus,

$$\hat{\sigma} = c_1 \cdot \sigma'_1 + \ldots + c_r \cdot \sigma'_r$$

is a feasible RLT-$t$ solution, since we have

$$\sum_{i=1}^{r} c_i = 1.$$

Let $f(\sigma)$ be the objective function value of solution $\sigma$. We have

$$f(\hat{\sigma}) = c_1 \cdot f(\sigma'_1) + \ldots + c_r \cdot f(\sigma'_r),$$

since $f$ is a linear function in terms of the $\vartheta^{(0)}$ and $\vartheta^{(1)}$ variables. Remark that only the vertices and the edges influence the objective function value of a QSAP. Thus, $f(\hat{\sigma}) = f(\sigma)$ must hold, because all $\vartheta^{(0)}$ and $\vartheta^{(1)}$ variables have the same values in $\hat{\sigma}$ and $\sigma$.

Since $\sigma$ is the optimal solution of the RLT-$t$ formulation and all solutions $\sigma'_s$ are also RLT-$t$ feasible, we know that the objective function value of $\sigma$ cannot be worse than the one of any $\sigma'_s$. Thus, we get

$$f(\sigma) \leq f(\sigma'_s) \quad \forall s \in \{1, ..., r\}.$$

It follows directly that

$$f(\sigma) = f(\sigma'_s) \quad \forall s \in \{1, ..., r\}$$

holds, since the solution $\hat{\sigma}$ is a linear combination of the solutions $\sigma'_1, ..., \sigma'_r$ and the objective function is also linear.

Every $\sigma'_s$ corresponds to a feasible solution of the QSAP. Hence, it follows that

$$\min \text{QSAP} \leq f(\sigma'_1) = f(\sigma) = \min \text{RLT-}t \leq \min \text{QSAP}$$

holds. This implies that the RLT-$t$ solution $\sigma$ is tight. $\qquad \square$

For the following results, we introduce the shorter notation

$$\vartheta^{(n-1)}_{C_i} = \vartheta^{(n-1)}_{i_1 j_1, ..., i_n j_n}$$

for the $\vartheta^{(n-1)}$ variable that corresponds to an $n$-clique $C_i = \{v_{i_1 j_1}, ..., v_{i_n j_n}\}$. In the next lemma, we show that every optimal RLT-$t$ solution of a QSAP of size $|M| \leq (t + 1)$ is tight. This lemma is a direct consequence of the RLT-$t$ constraints.

**Lemma 5.18** *A QSAP for which the RLT-t formulation is untight must have at least the size $|M| \geq (t + 2)$.*

**Proof:**

We know from Lemma 5.6 that the RLT-$t$ constraints enforce $(t + 1)$-cliques. Thus, the solution graph $G_{sol}$ of a solution $\sigma$ of the RLT-$t$ formulation for a QSAP with $|M| \leq (t + 1)$ rows consists of a union of $m$-cliques $C_1, ..., C_r$.

Here, the $\vartheta^{(0)}$ and $\vartheta^{(1)}$ variables fulfill the requirements of Lemma 5.17 with

$$\sigma(\vartheta_{ij}^{(0)}) = \sum_{\substack{s=1 \\ v_{ij} \in C_s}}^{r} \vartheta_{C_s}^{(m-1)} \quad \text{and}$$

$$\sigma(\vartheta_{ijkl}^{(1)}) = \sum_{\substack{s=1 \\ v_{ij}, v_{kl} \in C_s}}^{r} \vartheta_{C_s}^{(m-1)}.$$

Hence, we know from Lemma 5.17 that the solution $\sigma$ is tight. Thus, the RLT-$t$ formulation is tight for QSAPs of size $|M| \leq (t+1)$. This implies that the number of rows of a QSAP for which the RLT-$t$ formulation is untight must be greater than $(t+1)$.

$\square$

The following lemma presents a result that is needed for the main proof.

**Lemma 5.19** *Given are the sets of variables*

$$\{\nu_1^1, ..., \nu_n^1\}, \quad ..., \quad \{\nu_1^s, ..., \nu_n^s\} \ (\nu_i^j \in \mathbb{R}, \ \nu_i^j \geq 0)$$

*with $s < n$ and the $n$ inequalities*

$$\nu_i^1 + ... + \nu_i^s \geq \nu_0 \quad \forall i \in \{1, ..., n\}.$$

*For a set $S$ of real numbers, let $\max_2\{S\}$ be the second largest element of $S$. Then the inequality*

$$\max_2\{\nu_1^1, ..., \nu_n^1\} + ... + \max_2\{\nu_1^s, ..., \nu_n^s\} \geq \nu_0$$

*holds.*

**Proof:**

Suppose that there exist values for the $\nu_i^j$-variables that fulfill the requirements of the lemma and for which

$$\max_2\{\nu_1^1, ..., \nu_n^1\} + ... + \max_2\{\nu_1^s, ..., \nu_n^s\} < \nu_0$$

holds. Since we have $n$ inequalities of the form

$$\nu_i^1 + ... + \nu_i^s \geq \nu_0 \quad (i \in \{1, ..., n\})$$

and we regard the $s$ second largest elements

$$\max{}_2\{\nu_1^j, ..., \nu_n^j\} \quad (j \in \{1, ..., s\}),$$

there must exist a $k \in \{1, ..., n\}$ so that the set $\{\nu_k^1, ..., \nu_k^s\}$ contains no element $\nu_k^l$ ($l \in \{1, ..., s\}$) with $\nu_k^l > \nu_i^l$ for all $i \in \{1, ..., n\} \setminus \{k\}$. Thus, $\nu_k^l \leq \max_2\{\nu_1^l, ..., \nu_n^l\}$ holds for all $l \in \{1, ..., s\}$ and we get

$$\nu_0 \leq \nu_k^1 + ... + \nu_k^s \leq \max{}_2\{\nu_1^1, ..., \nu_n^1\} + ... + \max{}_2\{\nu_1^s, ..., \nu_n^s\} < \nu_0.$$

This contradicts the assumption.

□

With this result, we can prove the following lemma that improves our knowledge of tight RLT-$t$ formulations.

**Lemma 5.20** *Given are a fixed RLT level $t$ and a QSAP size $|M| = (t+2)$, $N = \{N_1, ..., N_m\}$ for which at least one $N_i$ has $n_i \leq t$ elements. The RLT-$t$ formulation is tight for all QSAPs of this size.*

**Proof:**

Given is a QSAP with $m = (t+2)$ rows that are without loss of generality ordered as a descending sequence with respect to the number of vertices ($n_1 \geq ... \geq n_m$) and for which $n_m \leq t$ holds.

We show that the requirements of Lemma 5.17 are fulfilled for every optimal solution $\sigma$ of this RLT-$t$ formulation. Thus, all possible solutions $\sigma$ are tight. From the RLT-$t$ constraints and Lemma 5.6 we know that in such a solution $\sigma$, the vertices and the edges of the first $(t+1)$ rows form a set of $(t+1)$-cliques $C_1, ..., C_r$. We denote the vertices of these cliques by

$$C_i = \{v_{(1)(j_1^i)}, ..., v_{(t+1)(j_{t+1}^i)}\} \quad (i \in \{1, ..., r\}).$$

An example of such a clique is shown in Figure 5.7 (left). To each of these cliques corresponds an RLT-$t$ variable $\vartheta_{C_s}^{(t)}$ with an assigned value greater than zero. From the RLT-$t$ constraints, we directly get

$$\sum_{s=1}^{r} \vartheta_{C_s}^{(t)} = \sum_{j_1 \in N_1} \cdots \sum_{j_{t+1} \in N_{t+1}} \vartheta_{(1)(j_1),...,(t+1)(j_{t+1})}^{(t)}$$

$$= \sum_{j_1 \in N_1} \cdots \sum_{j_t \in N_t} \vartheta_{(1)(j_1),...,(t)(j_t)}^{(t-1)}$$

$$= ...$$

$$= \sum_{j_1 \in N_1} \vartheta_{(1)(j_1)}^{(0)}$$

$$= 1.$$

We denote the vertices of row $(t + 2)$ by

$$row_{(t+2)} = \{v_{(t+2)(1)}, ..., v_{(t+2)(n_{t+2})}\}.$$

For every $i \in \{1, ..., r\}$, the $(t+1)$-clique $C_i$ contains $(t+1)$ different $t$-cliques $C_i^1, ..., C_i^{t+1}$. Such a clique $C_i^k$ is generated from $C_i$ by a removal of the vertex of row $k$. These cliques have the form

$$C_i^k = \{v_{(1)(j_1^i)}, ..., v_{(k-1)(j_{k-1}^i)}, v_{(k+1)(j_{k+1}^i)}, ..., v_{(t+1)(j_{t+1}^i)}\} = C_i \setminus \{v_{(k)(j_k^i)}\}.$$

Such a clique $C_i^k$ is shown in Figure 5.7 (right).

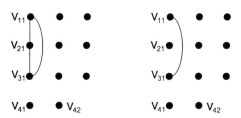

Figure 5.7: Examples of cliques $C_i$ (left) and $C_i^2$ (right) for $t = 2$.

For the RLT-$(t - 1)$ variables that correspond to these $t$-cliques, we get from the RLT-$t$ constraints the equation

$$\vartheta_{C_i^k}^{(t-1)} = \sum_{\substack{s=1 \\ C_i^k \subseteq C_s}}^{r} \vartheta_{C_s}^{(t)}.$$

From the RLT-$t$ constraints we know that there must exist $\vartheta^{(t)}$ variables with a value greater than zero for every set of $(t+1)$ rows. For each of these variables, there exists a $t+1$-clique in the solution graph of $\sigma$. This also holds for the sets of rows that contain row $(t+2)$. These $(t+1)$-cliques arise from a $t$-clique $C_i^k$ which is extended by a vertex of row $(t+2)$. We denote such a new $(t+1)$-clique by

$$C_{i,l}^k = \left\{ v_{(1)(j_1^i)}, ..., v_{(k-1)(j_{k-1}^i)}, v_{(k+1)(j_{k+1}^i)}, ..., v_{(t+1)(j_{t+1}^i)}, v_{(t+2)(l)} \right\}$$
$$= C_i^k \cup \{ v_{(t+2)(l)} \} \quad (k \in \{1, ..., t+1\},\ l \in \{1, ..., n_{t+2}\}).$$

Such a clique is presented in Figure 5.8 (left). With the corresponding variables $\vartheta_{C_{i,l}^k}^{(t)}$ and the RLT-$t$ constraints, we get the equation

$$\vartheta_{C_i^k}^{(t-1)} = \sum_{l=1}^{n_{t+2}} \vartheta_{C_{i,l}^k}^{(t)}. \tag{5.1}$$

For the next step, we need the $(t+2)$-clique that is generated by adding vertex $v_{(t+2)(l)}$ ($l \in \{1, ..., n_{t+2}\}$) to the $(t+1)$-clique $C_i$. We denote this new clique by

$$C_{i,l} = \left\{ v_{(1)(j_1)}, ..., v_{(t+1)(j_{t+1})}, v_{(t+2)(l)} \right\} = C_i \cup \{ v_{(t+2)(l)} \}.$$

Such a clique is shown in Figure 5.8 (right). We will show that the RLT-$t^{\leq}$

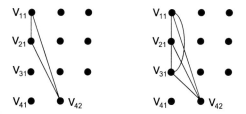

Figure 5.8: Examples of cliques $C_{i,2}^3$ (left) and $C_{i,2}$ (right) for $t = 2$.

constraints enforce that each $(t+1)$-clique $C_i$ is extended in the solution graph of $\sigma$ to one or several $(t+2)$-cliques $C_{i,l}$. Additionally, we will show that there are no other edges from the first $(t+1)$ rows to row $(t+2)$ than the ones that are part of these $(t+2)$-cliques.

From the RLT-$t$ constraints we know that the set of all $\vartheta_{C_i}^{(t)}$ variables uniquely defines the values for all RLT-$t'$ variables with $t' < t$. Therefore, we introduce for every $C' \subseteq C_i$ with $t' = |C'| + 1$ the new variable $\hat{\vartheta}_{C'}^{i,(t')}$. It defines the influence of the RLT-$t$ variable $\vartheta_{C_i}^{(t)}$ on the RLT-$t'$ variable $\vartheta_{C'}^{(t')}$. For each $t'$-clique $C'$, there must exist an extended $(t'+1)$-clique $C'_l$ that includes vertex $v_{(t+2)(l)}$ of row $(t+2)$ $(l \in \{1, ..., n_{t+2}\})$. We denote the influence of $\vartheta_{C_i}^{(t)}$ on the corresponding variable by $\hat{\vartheta}_{C'_l}^{i,(t'+1)}$. These influences on the RLT-$(t'+1)$ variables add up to the original RLT variable in the form

$$\vartheta_{C'_l}^{(t'+1)} = \sum_{\substack{s=1 \\ C' \subseteq C_s}}^{r} \hat{\vartheta}_{C'_l}^{s,(t'+1)}.$$

We know from the RLT-$t$ constraints that these new variables that correspond to cliques $C' \subseteq C_i$ satisfy

$$\sum_{l=1}^{n_{t+2}} \hat{\vartheta}_{C'_l}^{i,(t'+1)} = \hat{\vartheta}_{C'}^{i,(t')} = \vartheta_{C_i}^{(t)} \tag{5.2}$$

for all $i \in \{1, ..., r\}$.

We regard the 2-clique $C'_l = \{v_{(k)(j_k^i)}, v_{(t+2)(l)}\}$ for the following analysis. For this edge, we define

$$\hat{\lambda}_{C_i,l} = \min_{k \in \{1,...,t+1\}} \left\{ \hat{\vartheta}_{\{v_{(k)(j_k^i)}, v_{(t+2)(l)}\}}^{i,(1)} \right\}$$

as the lowest influence of $\vartheta_{C_i}^{(t)}$ on an edge that is between a vertex of $C_i$ and the vertex $v_{(t+2)(l)}$. From the inequality

$$\hat{\vartheta}_{\{v_{(k)(j_k^i)}, v_{(t+2)(l)}\}}^{i,(1)} \geq \max_{u \in \{1,...,t+1\} \setminus \{k\}} \left\{ \hat{\vartheta}_{C_{i,l}^u}^{i,(t)} \right\},$$

we get

$$\hat{\lambda}_{C_i,l} \geq \min_{k \in \{1,...,t+1\}} \left\{ \max_{u \in \{1,...,t+1\} \setminus \{k\}} \left\{ \hat{\vartheta}_{C_{i,l}^u}^{i,(t)} \right\} \right\} = \max_2 \left\{ \hat{\vartheta}_{C_{i,l}^1}^{i,(t)}, ..., \hat{\vartheta}_{C_{i,l}^{t+1}}^{i,(t)} \right\}.$$

With (5.2), which we use in the form

$$\sum_{l=1}^{n_{t+2}} \hat{\vartheta}_{C_{i,l}^u}^{i,(t)} = \vartheta_{C_i}^{(t)},$$

the requirements of Lemma 5.19 are fulfilled and we get

$$\sum_{l=1}^{n_{t+2}} \hat{\lambda}_{C_i,l} \geq \sum_{l=1}^{n_{t+2}} \max_2 \left\{ \hat{\vartheta}_{C_{i,l}^1}^{i,(t)}, ..., \hat{\vartheta}_{C_{i,l}^{t+1}}^{i,(t)} \right\} \geq \vartheta_{C_i}^{(t)}.$$

This means that for every $(t+1)$-clique $C_i$, there exist $(t+2)$-cliques in the solution graph that are extensions of $C_i$. The $\vartheta^{(1)}$ variables with minimal value that extend such a clique $C_i$ to a larger clique $C_{i,l}$ sum up at least to the value $\vartheta^{(t)}_{C_i}$.

We still have to show that no further edges than the ones of these cliques $C_{i,l}$ are in the solution graph of $\sigma$ and that the requirements of Lemma 5.17 are fulfilled. By combining the previous results, we get

$$
\begin{aligned}
1 &= \sum_{s=1}^{r} \vartheta^{(t)}_{C_s} \\
&\leq \sum_{s=1}^{r} \sum_{l=1}^{n_{t+2}} \hat{\lambda}_{C_{s,l}} \\
&\leq \sum_{s=1}^{r} \sum_{l=1}^{n_{t+2}} \hat{\vartheta}^{s,(1)}_{\{v_{(k)(j_k^s)}, v_{(t+2)(l)}\}} \quad (k \in \{1, ..., t+1\}) \\
&= \sum_{j=1}^{n_k} \sum_{l=1}^{n_{t+2}} \sum_{\substack{s=1 \\ v_{(k)(j)} \in C_s}}^{r} \hat{\vartheta}^{s,(1)}_{\{v_{(k)(j)}, v_{(t+2)(l)}\}} \\
&= \sum_{j=1}^{n_k} \sum_{l=1}^{n_{t+2}} \vartheta^{(1)}_{\{v_{(k)(j)}, v_{(t+2)(l)}\}} \\
&= 1.
\end{aligned}
$$

This implies that

$$
\hat{\lambda}_{C_{s,l}} = \hat{\vartheta}^{s,(1)}_{\{v_{(k)(j_k^s)}, v_{(t+2)(l)}\}}
$$

holds for all $k \in \{1, ..., t+1\}$. Hence, the values $\hat{\lambda}_{C_{s,l}}$ uniquely define all RLT-1 variables for edges between vertices of $C_s$ and vertices of row $(t+2)$ for all $(t+1)$-cliques $C_s$.

The values $\hat{\lambda}_{C_{i,l}}$ $(i \in \{1, ..., r\}, l \in \{1, ..., n_{t+2}\})$ that correspond to the $(t+2)$-cliques $C_{i,l}$ fulfill the requirements of Lemma 5.17. This can be shown as follows:

For edges $e_{ijpq}$ with $i < p < (t + 2)$, we have

$$
\begin{aligned}
\vartheta_{ij,pq}^{(1)} &= \sum_{\substack{s=1 \\ v_{(i)(j)}, v_{(p)(q)} \in C_s}}^{r} \vartheta_{C_s}^{(t)} \\
&= \sum_{\substack{s=1 \\ v_{(i)(j)}, v_{(p)(q)} \in C_s}}^{r} \hat{\vartheta}_{\{v_{(i)(j)}\}}^{s,(0)} \\
&= \sum_{\substack{s=1 \\ v_{(i)(j)}, v_{(p)(q)} \in C_s}}^{r} \sum_{l=1}^{n_{t+2}} \hat{\vartheta}_{\{v_{(i)(j)}, v_{(t+2)(l)}\}}^{s,(1)} \\
&= \sum_{\substack{s=1 \\ v_{(i)(j)}, v_{(p)(q)} \in C_s}}^{r} \sum_{l=1}^{n_{t+2}} \hat{\lambda}_{C_{s,l}} \\
&= \sum_{l=1}^{n_{t+2}} \sum_{\substack{s=1 \\ v_{(i)(j)}, v_{(p)(q)} \in C_{s,l}}}^{r} \hat{\lambda}_{C_{s,l}}
\end{aligned}
$$

and for edges $e_{(i)(j)(t+2)(l)}$, we have

$$
\vartheta_{(i)(j),(t+2)(l)}^{(1)} = \sum_{\substack{s=1 \\ v_{ij} \in C_s}}^{r} \hat{\vartheta}_{\{v_{ij}, v_{(t+2)(l)}\}}^{s,(1)} = \sum_{\substack{s=1 \\ v_{ij} \in C_{s,l}}}^{r} \hat{\lambda}_{C_{s,l}} .
$$

The $\vartheta^{(0)}$ variables for vertices of the first $(t + 1)$ rows fulfill

$$
\vartheta_{ij}^{(0)} = \sum_{l=1}^{n_{t+2}} \vartheta_{(i)(j),(t+2)(l)}^{(1)} = \sum_{l=1}^{n_{t+2}} \sum_{\substack{s=1 \\ v_{ij} \in C_s}}^{r} \hat{\vartheta}_{(i)(j),(t+2)(l)}^{s,(1)} = \sum_{l=1}^{n_{t+2}} \sum_{\substack{s=1 \\ v_{ij} \in C_{s,l}}}^{r} \hat{\lambda}_{C_{s,l}}
$$

and for vertices of row $(t + 2)$ we have

$$
\vartheta_{(t+2)(l)}^{(0)} = \sum_{j=1}^{n_1} \vartheta_{(1)(j),(t+2)(l)}^{(1)} = \sum_{j=1}^{n_1} \sum_{\substack{s=1 \\ v_{ij} \in C_s}}^{r} \hat{\vartheta}_{(1)(j),(t+2)(l)}^{s,(1)} = \sum_{s=1}^{r} \hat{\lambda}_{C_{s,l}} .
$$

Thus, we finally get from Lemma 5.17 that the optimal RLT-$t$ solution $\sigma$ is tight. Since this result is independently of the objective function, the RLT-$t$ formulation is tight for all QSAPs of the given problem size.

$\square$

In this proof we have up to now shown that the RLT-$t$ formulation is tight for all QSAPs of size $m < (t+2)$ and for the QSAPs of size $m = (t+2)$ with at least one row with $n_i < (t+1)$ vertices. We extend this result in the following lemma, where we add additional rows with less than $(t + 1)$ vertices to the QSAP formulation from Lemma 5.20. We show that the RLT-$t$ formulation is still tight for this extended QSAP.

**Lemma 5.21** *Given are a fixed RLT level $t$ and a QSAP with $m = (t + 2)$ rows and at least one row $i$ with $n_i < (t + 1)$ vertices. Extending the QSAP size by rows with less than $(t + 1)$ vertices results in a new QSAP size with $m' > m$ rows for which the corresponding RLT-t formulation is tight.*

**Proof:**

From the proof of Lemma 5.20, we know that the RLT-$t$ formulation is tight for the QSAP that consists of the first $(t + 2)$ rows. Additionally, we know that the graph $G_{sol}$ that corresponds to an optimal solution of the RLT-$t$ fomulation consists of a union of $(t+2)$-cliques $C_1, ..., C_r$ with a corresponding value $\hat{\lambda}_{C_i}$ for clique $C_i$. For such a clique $C_i$, we introduce the notation

$$C_i = \{v_{(1)(j_1^i)}, ..., v_{(t+2)(j_{t+2}^i)}\}.$$

We sort the first $m = (t + 2)$ rows in descending order. Thus, row $(t + 2)$ has less than $(t + 1)$ vertices. We first consider the extension of the QSAP by only one additional row. The proof for this case is similar to the one of Lemma 5.20. Therefore, we will focus in this proof on the main ideas that are different from the previous proof.

If we remove the vertex of row $k$ from clique $C_i$, we get a $(t + 1)$-clique. We denote this new clique by $C_i^k$. If we consider the set of the first $(t + 2)$ rows, remove one of the rows and replace it with row $(t+3)$, the requirements of Lemma 5.20 are fulfilled since $n_{t+3} < t+1$ holds. Thus, any RLT-$t$ feasible solution for these $(t + 2)$ rows also consists of a union of $(t + 2)$-cliques. We denote these possible cliques by $C_{i,q}^k$ for $q \in \{1, ..., n_{t+3}\}$.

Similar to the proof of the previous lemma, we introduce the influence of clique $C_i$ on all edges of an RLT-$t'$ variable ($t' < t + 1$) that corresponds to a $(t+1)$-clique $C'$ by $\hat{\lambda}_{C'}^i$. According to this, we introduce by $\hat{\lambda}_{C'_q}^i$ the influence of clique $C_i$ on all edges of an RLT-$(t' + 1)$ variable that corresponds to a $(t + 2)$-clique $C'_q$ that contains $(t + 1)$ vertices of $C_i$ and vertex $v_{(t+3)(q)}$. For these new variables, we have

$$\sum_{q=1}^{n_{t+3}} \hat{\lambda}_{C'_q}^i = \hat{\lambda}_{C'}^i = \hat{\lambda}_{C_i}. \tag{5.3}$$

For the edges between a vertex $v_{(k)(j_k^i)}$ of a $(t+2)$-clique $C_i$ and vertex $v_{(t+3)(q)}$, we define $\rho_{C_{i,q}}$ as the smallest of these edge values $\hat{\lambda}_{\{v_{(k)(j_k^i)}, v_{(t+3)(q)}\}}^i$. For these new variables, the inequality

$$\rho_{C_{i,q}} \geq \min_{j \in \{1,...,t+2\}} \left\{ \max_{k \in \{1,...,t+2\} \setminus \{j\}} \left\{ \hat{\lambda}_{C_{i,q}^k}^i \right\} \right\} = \max_2 \left\{ \hat{\lambda}_{C_{i,q}^1,...,C_{i,q}^{t+2}}^i \right\}$$

holds. Combined with equation (5.3) in the form

$$\sum_{q=1}^{n_{t+3}} \hat{\lambda}_{C_{i,q}^k}^i = \hat{\lambda}_{C_i} \quad \forall k \in \{1,...,t+2\},$$

the requirements of Lemma 5.19 are fulfilled and we get

$$\sum_{q=1}^{n_{t+3}} \rho_{C_{i,q}} \geq \hat{\lambda}_{C_i}.$$

With this inequality, we have

$$1 = \sum_{s=1}^{r} \hat{\lambda}_{C_s}$$

$$\leq \sum_{s=1}^{r} \sum_{q=1}^{n_{t+3}} \rho_{C_{s,q}}$$

$$\leq \sum_{s=1}^{r} \sum_{q=1}^{n_{t+3}} \hat{\lambda}_{\{v_{(k)(j_k^s)}, v_{(t+3)(q)}\}}^s \quad (k \in \{1,...,t+2\})$$

$$= \sum_{j_k=1}^{n_k} \sum_{q=1}^{n_{t+3}} \sum_{\substack{s=1 \\ v_{(k)(j)} \in C_s}}^{r} \hat{\lambda}_{\{v_{(k)(j_k)}, v_{(t+3)(q)}\}}^s$$

$$= \sum_{j=1}^{n_k} \sum_{q=1}^{n_{t+3}} \vartheta_{(k)(j),(t+3)(q)}^{(1)}$$

$$= 1.$$

This implies

$$\rho_{C_{s,q}} = \hat{\lambda}_{\{v_{(k)(j_k^s)}, v_{(t+3)(q)}\}}^s$$

for all $k \in \{1, ..., t+2\}$.

Since the $\hat{\lambda}$ variables uniquely define the RLT-1 variables that correspond to edges of $G_{QSAP}$, we get the equations

$$\vartheta^{(1)}_{ijkl} = \sum_{s=1}^{r} \sum_{\substack{q=1 \\ v_{ij}, v_{kl} \in C_{s,q}}}^{n_{t+3}} \rho_{C_{s,q}}$$

and

$$\vartheta^{(0)}_{ij} = \sum_{s=1}^{r} \sum_{\substack{q=1 \\ v_{ij} \in C_{s,q}}}^{n_{t+3}} \rho_{C_{s,q}}.$$

Hence, the requirements of Lemma 5.17 are fulfilled and we get that the RLT-$t$ formulation is tight for this extended QSAP with $m = (t+3)$ rows.

The same argumentation can be done for further extensions of this QSAP with rows with less than $(t+1)$ vertices. The proofs for these extensions use the previous step of the here presented proof instead of being based on the result of Lemma 5.20. This proves the lemma by induction.

$\square$

By combining the previous lemmas, we can prove another lemma which shows that the RLT-$\gamma(m, N)$ formulation is tight.

**Lemma 5.22** *For a QSAP size $|M| = m$ and $N = \{N_1, ..., N_m\}$, the RLT-$\gamma(m, N)$ formulation is tight, independently of the objective function.*

**Proof:**

Given is a QSAP of size $|M| = m$ and we assume without loss of generality that the row sizes $n_i$ $(i \in M)$ are sorted in descending order. We know from Lemma 5.20 that a graph with $m = (t+2)$ rows and at least one row $i$ with $n_i < (t+1)$ vertices that satisfies all RLT-$t^{\leq}$ constraints corresponds to a tight solution.

Thus, the RLT-$\gamma(m, N)$ formulation is tight for the QSAP size that is defined by the first $m' = (\gamma(m, N) + 2)$ rows $N_1, ..., N_{m'}$ of the given QSAP.

From Lemma 5.21, we know that adding rows with less than $\gamma(m, N) + 1$ vertices to this reduced QSAP does not effect the tightness of the formulation. Additionally, we know from Definition 5.15 that

$$\gamma(m, N) = \min\{s \in \mathbb{N} \mid s + 1 > n_{i_{s+2}}\} \geq n_{i'}$$

holds for all $i' > m'$. Hence, the requirements of Lemma 5.21 are fulfilled and the RLT-$\gamma(m, N)$ formulation is tight for the initial QSAP.

$\square$

With the result of the previous lemma, we can construct for each QSAP size an RLT level $t = \gamma(m, N)$ for which the RLT-$t$ formulation is tight. For the proof of Theorem 5.16, we still have to show that the RLT-$(\gamma(m, N) - 1)$ formulation is not tight.

**Lemma 5.23** *For a given QSAP size $|M| = m$ and $N = \{N_1, ..., N_m\}$, the RLT-$t'$ formulation with $t' \leq (\gamma(m, N) - 1)$ is not tight.*

**Proof:**

In the proof of Lemma 5.14, we constructed a QSAP of size $m = (t + 2)$ and $n_i = (t + 1)$ for all $i \in M$ for which the optimal RLT-$t$ solution is not tight. We will use the same idea in the following argumentation.

For a given QSAP of a certain size, we know from the definition of $\gamma(m, N)$ that this QSAP has $(t' + 2)$ rows with at least $(t' + 1)$ vertices in each row. Following the proof of Lemma 5.14, we can construct a QSAP for which the optimal RLT-$t'$ solution is not tight. For this, we embed the graph $G_{RLT}^{t'} = (V_{RLT}^{t'}, E_{RLT}^{t'})$ in the $(t'+2)$ rows with the highest number of vertices. These rows have the indices $i_1, ..., i_{t'+2}$. An objective function for which the RLT-$t'$ formulation generates an untight optimal solution can be constructed by

$$b_{ij} = 0 \quad \forall i \in M, \ \forall j \in N_i \quad \text{and}$$

$$c_{ijkl} = \begin{cases} 0, & \text{if } i \notin \{i_1, ..., i_{t'+2}\}, \\ 0, & \text{if } i \in \{i_1, ..., i_{t'+2}\}, k \notin \{i_1, ..., i_{t'+2}\}, \\ 0, & \text{if } e_{ijkl} \in E_{RLT}^{t'}, \\ 1, & \text{if } i, k \in \{i_1, ..., i_{t'+2}\}, i < k, \ e_{ijkl} \notin E_{RLT}^{t'}. \end{cases}$$

The optimal solution of the RLT-$t'$ formulation is zero. In contrast to this, the optimal solution value of the QSAP is one, since at least one edge $e_{ijkl}$ with a coefficient $c_{ijkl} = 1$ must be chosen to construct a $(t'+2)$-clique in the rows $i_1, ..., i_{t'+2}$. Thus, the lower bound gained by the RLT-$t'$ formulation is not tight.

Therefore, the RLT-$t'$ formulation is not tight for the QSAPs of the given size, since we can construct a QSAP for which the optimal RLT-$t'$ solution is not tight.

□

Lemma 5.22 and Lemma 5.23 combined finally prove Theorem 5.16. We now present an interpretation of the results of Theorem 5.16 on graphs. If the rows of a QSAP-graph are ordered concerning their size in descending order, the largest $(t+2) \times (t+1)$ rectangle ($(t+2)$ rows, $(t+1)$ columns) that can be embedded in this graph defines the smallest level $(t+1)$ for which the RLT-$(t+1)$ formulation is tight for the given QSAP size. Such a rectangular visualization is presented in Figure 5.9.

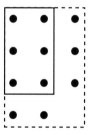

Figure 5.9: Vertex structure of a QSAP with a maximal fitting $(t+2) \times (t+1)$ rectangle that can be embedded in the problem graph. The continuous rectangle represents the size of the graph $G^1_{RLT}$. The dashed lines shows the smallest size of a graph $G^t_{RLT}$ that does not fit into the QSAP graph. Since $G^2_{RLT}$ cannot be a subgraph of a QSAP of this size, the RLT-2 formulation is tight for all QSAPs of this size.

We can use the results of this chapter to get a better insight in the theoretic structure of the RLT. The results improve the currently known trivial level $t = m - 1$ that provides a tight formulation for a QSAP size. Especially for the TTSP, where the problem structure often contains rows $N_i$ with $n_i = 1$, this knowledge can be helpful. Furthermore, we will use the gained knowledge for a new type of algorithm that successively removes the problem structures $G^t_{RLT}$ by partially increasing the RLT level. This new algorithm is presented in Section 5.7.

### 5.6.3 Polyhedral Interpretation of the Results

In Theorem 5.16, we have shown that the RLT-$t$ formulation of a QSAP of size $|M| = m$ and $N = \{N_1, ..., N_m\}$ is tight if and only if $t \geq \gamma(m, N)$ holds. This can also be interpreted as a polyhedral result.

**Theorem 5.24** *The projection of the polytope that is defined by the RLT-$\gamma(m, N)$ formulation to the x-variables and the y-variables of the LP formulation characterizes the convex hull of the solutions of the QSAP.*

**Proof:**

To prove the theorem, we have to show that every extreme point of the projected RLT-$\gamma(m, N)$ polytope corresponds to a solution of the QSAP. In the proofs of Lemma 5.20 and Lemma 5.21, we have shown that every solution of the RLT-$\gamma(m, N)$ formulation consists only of a union of $m$-cliques. This structure implies that the projection to the $x$-variables and the $y$-variables of each solution of the RLT-$\gamma(m, N)$ formulation is a convex combination of solutions of the QSAP. Thus, every point in the projected polytope is a solution of the QSAP or a convex combination of solutions. This implies that every extreme point of the polytope is a solution of the QSAP, which proves the theorem.

$\square$

With the presented results, we can improve the known statement

$$P_0 = P_{RLT-1} \supseteq P_{RLT-2} \supseteq ... \supseteq P_{RLT-(m-1)} = conv(X)$$

for the projections of the RLT-polyhedrons to the $x$ and $y$ variables to

$$P_0 \supset P_{RLT-1} \supset ... \supset P_{RLT-\gamma(m,N)} = ... = conv(X),$$

where $conv(X)$ is the convex hull of the feasible solutions of the initial QSAP and where we have proper subsets.

### 5.6.4 Tight RLT-Formulations for the QAP

We now make similar considerations of tight formulations for QAPs, whose RLT structure was introduced in 5.3. Due to the higher number of constraints, the RLT formulations become tighter at earlier RLT levels. Nevertheless, finding a general graph structure that satisfies the RLT constraints but that contains no $m$-clique is a hard task.

We start again with the general problem structure for a problem of size $|M| = m$. The corresponding QAP graph $G_{QAP}$ was introduced in Definition 3.11:

$$G_{QAP} = (V_{QSAP}, E_{QSAP})$$
$$V_{QAP} = \{v_{ij} \mid i, j \in M\},$$
$$E_{QAP} = \{e_{ijkl} = \{v_{ij}, v_{kl}\} \mid i, j, k, l \in M, i < k, j \neq l\}.$$

Such a graph for $|M| = 4$ can be seen in Figure 5.10.

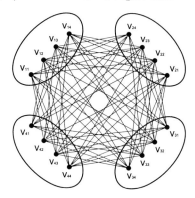

Figure 5.10: Graphical representation $G_{QSAP}$ of a QAP with $m = 4$.

For the RLT-1 formulation, the minimal graph (in terms of number of vertices) that satisfies the RLT-$1^{\leq}$ constraints and that contains no $m$-clique is presented in Figure 5.11. This graph has the size $|M| = 4$ and the structure

$$\widetilde{G}_{RLT}^1 = (\widetilde{V}_{RLT}^1, \widetilde{E}_{RLT}^1),$$
$$\widetilde{V}_{RLT}^1 = \{v_{ij} \in V_{QAP}\},$$
$$\widetilde{E}_{RLT}^1 = \{e_{ijkl} \in E_{QAP} \; : \; l \not\equiv (j + k - i) \mod 4\}.$$

The corresponding feasible variable assignment that is needed for the RLT solution is given by

$$x_{ij} = \frac{1}{4} \quad \forall i \in M, \; \forall j \in N_i,$$
$$y_{ijkl} = \begin{cases} \frac{1}{8}, & \text{if } e_{ijkl} \in \widetilde{E}_{RLT}^1, \\ 0, & \text{else.} \end{cases}$$

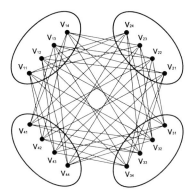

Figure 5.11: Graph $\widetilde{G}^1_{RLT}$ with $m = 4$ that contains no 4-clique but that fulfills all RLT-$1^\le$ constraints.

The graph $\widetilde{G}^1_{RLT}$ contains no 4-clique, but we still have to show that it fulfills all RLT-$1^\le$ constraints. This is done in the following proposition.

**Proposition 5.25** *The graph $\widetilde{G}^1_{RLT}$ satisfies all RLT-$1^\le$ constraints.*

**Proof:**

Since $x_{ij} = \frac{1}{4}$ holds for all $i, j \in M$, we directly see from

$$\sum_{i \in M} x_{ij} = 4 \cdot \frac{1}{4} = 1 \quad \forall j \in M \quad \text{and}$$

$$\sum_{j \in M} x_{ij} = 4 \cdot \frac{1}{4} = 1 \quad \forall i \in M$$

that the RLT-0 constraints hold.

For the first type of RLT-1 constraints, namely

$$\sum_{l \in M \setminus \{j\}} y_{ijkl} = x_{ij},$$

we have to show that each vertex $v_{ij}$ has exactly two adjacent vertices $v_{kl_1}, v_{kl_2}$ for every $k \in M \setminus \{i\}$. From the definition of $\widetilde{G}^1_{RLT}$, we know that for fixed $i, j, k \in M$ $(i \ne k)$, $\{v_{ij}, v_{kl}\} \in \widetilde{E}^1_{RLT}$ holds if and only if $l \not\equiv j \mod 4$ and $l \not\equiv (j+k-i) \mod 4$ hold. Since $k \not\equiv i \mod 4$ holds, we have $j \not\equiv (j+k-i)$

mod 4. Thus, two of the four possible vertices $v_{k1}, v_{k2}, v_{k3}, v_{k4}$ are adjacent to $v_{ij}$ and we get

$$\sum_{l \in M \setminus \{j\}} y_{ijkl} = 2 \cdot \frac{1}{8} = \frac{1}{4} = x_{ij}.$$

To show that the second type of RLT-1 constraint

$$\sum_{k \in M \setminus \{i\}} y_{ijkl} = x_{ij}$$

also holds for $\widetilde{G}^1_{RLT}$, we prove that each vertex $v_{ij}$ has exactly two adjacent vertices of the form $v_{k_1 l}, v_{k_2 l}$ for every $l \in M \setminus \{j\}$. For fixed $i, j, l \in M$ ($j \neq l$), we directly get that $v_{kl}$ is adjacent to $v_{ij}$ if and only if $k \neq i$ and $k \not\equiv (l - j + i) \mod 4$. Since $l \not\equiv j \mod 4$ holds, we have $i \not\equiv (l - j + i) \mod 4$. Thus, exactly two of the four vertices $v_{1l}, v_{2l}, v_{3l}, v_{4l}$ are adjacent to $v_{ij}$ and we get

$$\sum_{k \in M \setminus \{i\}} y_{ijkl} = 2 \cdot \frac{1}{8} = \frac{1}{4} = x_{ij}.$$

Hence, all RLT-$1^{\leq}$ constraints are satisfied.

$\square$

The following proposition presents a first result for RLT-1 feasible solutions of QAPs with less than four rows. Here, we show that the RLT-1 formulation implies $m$-cliques, which is a first indicator of a tight formulation. Note that the approach from Lemma 5.19 that was used for the QSAP cannot be applied to the QAP.

**Proposition 5.26** *Every solution graph of an RLT-1 feasible solution of a QAP with $m < 4$ rows contains an $m$-clique.*

**Proof:**

For $m = 2$, the RLT-1 constraints force the corresponding graph to have at least one edge. Thus, the graph contains a 2-clique.

We consider now the QAP-graph $G_{QAP}$ with $m = 3$. Suppose that there exists a subgraph $\hat{G} = \{\hat{V}, \hat{E}\}$ of $G_{QAP}$ that corresponds to an RLT-1 solution and that contains no 3-clique. We know from the RLT-$1^{\leq}$ constraints that each vertex $v_{ij}$ that is adjacent to an edge in $\hat{G}$ is incident to at least one vertex $v_{kl}$ for all $k \neq i$ and at least one vertex $v_{st}$ for all $t \neq j$.

Since $\hat{G}$ contains no 3-clique, the graph must have at least one edge $e_{ijkl}$ less than $G_{QAP}$ with $x_{ij} > 0$ and $x_{kl} > 0$. Without loss of generality, let this edge be $\{v_{11}, v_{22}\} \notin \hat{E}$. It follows directly that $\{v_{11}, v_{23}\} \in \hat{E}$, $\{v_{11}, v_{32}\} \in \hat{E}$, $\{v_{13}, v_{22}\} \in \hat{E}$ and $\{v_{22}, v_{31}\} \in \hat{E}$ hold (cf. Figure 5.12).

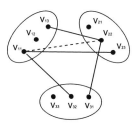

Figure 5.12: Construction step for $\hat{G}$. The dashed line represents an edge that cannot an element of $\hat{E}$.

Thus, it is easy to conclude that $\{v_{23}, v_{32}\} \notin \hat{E}$ holds (this edge would cause a 3-clique in $\hat{G}$). Hence, $\{v_{13}, v_{32}\} \in \hat{E}$ holds. Additionally, by the same argumentation, we get $\{v_{13}, v_{31}\} \notin \hat{E}$ and $\{v_{13}, v_{21}\} \in \hat{E}$.

With $\{v_{13}, v_{21}\} \in \hat{E}$ and $\{v_{13}, v_{32}\} \in \hat{E}$, we get $\{v_{21}, v_{32}\} \notin \hat{E}$ (cf. Figure 5.13). Hence, we have $\{v_{21}, v_{32}\} \notin \hat{E}$ and $\{v_{23}, v_{32}\} \notin \hat{E}$ which contradicts the RLT-1 constraint

$$\sum_{j=1}^{3} y_{2j32} = x_{32}$$

that corresponds to these vertices.

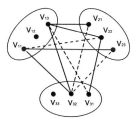

Figure 5.13: Construction step for $\hat{G}$. The dashed lines represent vertex pairs where no edges are allowed.

Therefore, such a graph $\hat{G}$ cannot exist. Hence, every solution graph of an RLT-1 feasible solution of a QAP of size $m \leq 3$ contains an $m$-clique.

$\square$

The previous lemma presents a first result for the tightness of the RLT for QAPs. But this topic is still an open field of research, where the RLT level that generates tight formulations for a certain QAP size is unknown.

## 5.7   RLT-1.5, a new Algorithmic Approach

The previous sections contain the concept of different RLT levels, an interpretation of the RLT formulations on graphs and new mathematical results for tight formulations. For the QSAP, the solutions of the RLT-1 formulation and the solutions of the RLT-2 formulation are analyzed. For the problem sizes that are regarded in this thesis, it can be seen that the RLT-1 formulation is able to compute lower bounds quickly, but the gap size is in most cases too high for practical purposes. The RLT-2 formulation, on the other hand, generates better lower bounds, but the time that is needed to solve the formulation is too long for bigger problem sizes. Thus, it stands to reason to search for a strategy that combines both, the speed of the RLT-1 formulation and the better solution quality of the RLT-2 formulation.

To achieve this combined algorithm that we call RLT-1.5 approach, we start by solving the RLT-1 formulation that is equivalent to the standard LP formulation of the QSAP. We analyze the solution graph of the optimal RLT-1 solution $\sigma_1$ for occurrences of $G^1_{RLT}$ or bigger graphs that are responsible for untight solutions. This is done by searching for adjacent vertices $v_{ij}$ and $v_{kl}$ $(i < k)$ that form no triangle with a third row $p$ with $i \neq p \neq k$ (there exists no vertex $v_{pq}$ in this row that is adjacent to both $v_{ij}$ and $v_{kl}$). Such a pair of vertices is a necessary condition for the existence of a graph $G^1_{RLT}$ that is responsible for an untight formulation.

For the found row triples $i, k, p$, we introduce new RLT-2 variables

$$z_{ijklpq} \geq 0 \quad \forall j \in N_i, \ \forall l \in N_k, \ \forall q \in N_p$$

and new RLT-2 constraints

$$\sum_{j=1}^{n_k} z_{ijklpq} = y_{klpq} \quad \forall l \in N_k, \ \forall q \in N_p,$$

$$\sum_{l=1}^{n_k} z_{ijklpq} = y_{ijpq} \quad \forall j \in N_i, \ \forall q \in N_p,$$

$$\sum_{q=1}^{n_k} z_{ijklpq} = y_{ijkl} \quad \forall j \in N_i, \ \forall l \in N_k$$

and add them to the RLT-1 formulation. These new constraints eliminate the possibility of an occurrence of the problem structure $G^1_{RLT}$ within these three rows. The solutions of the resulting linear program cannot contain edges between the rows $i$, $k$ and $l$ that form no triangle with the third row.

The generated bounds of the new formulation cannot be worse than the bounds of the RLT-1 formulation and we get in most cases an improvement by the new formulation. By optimizing the new formulation, we get the optimal solution $\sigma_2$, which we analyze again as we did before with the RLT-1 solution. In the corresponding solution graph, we search again for edges that form no triangles with a third row. Introducing new RLT-2 variables and RLT-2 constraints generates once again a tighter polyhedral formulation.

Repeating these steps results in a stepwise process that provides better lower bounds for a given QSAP. The algorithm is related to the theory of cutting plane algorithms, but the additional variables that are added to the formulation in each step makes it a more abstract approach. With this approach, we get the desired intermediate steps for the polyhedral formulation. These steps are of the form

$$P_{\text{RLT-1}} \supseteq P^1_{\text{RLT-1.5}} \supseteq \cdots \supseteq P^n_{\text{RLT-1.5}} \supseteq P_{\text{RLT-2}},$$

where $P^s_{\text{RLT-1.5}}$ describes the projection of the polytope that is generated after $s$ improvement steps to the space of the original $x$ and $y$ variables. The flowchart of the algorithm is presented in Figure 5.14.

Using modern LP-solvers like Cplex 11.2 for this kind of stepwise optimization algorithm has the advantage that the solvers store information about the last iteration step and use it in an intelligent way for further calculations. Thus, the new RLT-1.5 formulations can be solved much faster in the given context than if the formulations are solved anew.

The RLT-1.5 algorithm is able to calculate promising lower bounds quickly. The advantage lies in the good convergence behavior in the early phase of the algorithm. If it comes to the calculation of best lower bounds, the algorithm is not a good choice, since it needs in most cases more time than the

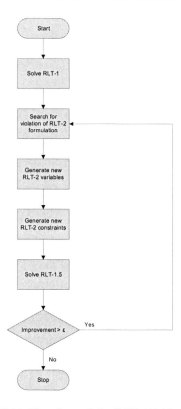

Figure 5.14: Flowchart of the RLT-1.5 Algorithm.

RLT-2 formulation. This is the case, since the solvers can directly use all cutting planes at the beginning for the RLT-2 formulation which gives it an advantage. An example of the convergence behavior of the RLT-1.5 formulation compared to the RLT-2 approach is presented in Figure 6.1. Here it can be seen that the RLT-1.5 approach is remarkably faster than the RLT-2 approach.

A comparison of the RLT-1.5 formulation to the RLT-1 approach is not required. The RLT-1.5 algorithm has the same behavior at the beginning, since it starts by solving the RLT-1 formulation. Detailed results of our RLT-1.5 approach are presented in Section 6.3.4.

The RLT approaches are not only used as a formulation that is solved via LP solvers. The technique also gives starting points for different algorithmic approaches. Here, the new RLT-1.5 formulations can help to get better formulations that represent a good trade-off between solution quality and computation time. Such an extended approach that does not need an LP solver to generate bounds for the QSAP is presented in the following section.

## 5.8 A Reduction Technique for the QSAP

A different approach that generates good lower bounds for the QSAP is to "reduce" the problem. This is done by extracting a constant summand $a'$ from the objective function $f(x)$ so that the general structure of the function is kept. This means that the new objective function $g(x) = a' + f'(x)$ still returns the same solution values for the set of all feasible assignments which means

$$f(x) = a' + f'(x) \quad \forall x \in \mathcal{A}.$$

The new function $f'(x)$ is created so that it has the structure

$$f'(x) = a' + \sum_{i=1}^{m} \sum_{j=1}^{n_i} b'_{ij} x_{ij} + \sum_{i=1}^{m-1} \sum_{k=i+1}^{m} \sum_{j=1}^{n_i} \sum_{l=1}^{n_k} c'_{ijkl} x_{ij} x_{kl}$$

with only nonnegative coefficients $b'_{ij}$ and $c'_{ijkl}$. The extracted constant value $a'$ can be used as a lower bound for the QSAP, since $f'(x) \geq 0$ holds for all $x \in \mathcal{A}$. Billionet and Elloumi present such a reduction approach for the QSAP in [5].

The main ideas of this technique are quite similar to the ones that the RLT approach is based on. Hence, some ideas of the previous section can also be applied to this reduction technique.

This section introduces the basic structures that are needed for the reduction approach. Additionally, some new theoretic results are presented and an algorithm that is based on these results is introduced. This approach is promising for the TTSP, since the objective functions for this type of problem have only nonnegative coefficients in the initial objective function $f(x)$.

## 5.8.1   Basic Definitions

In this section, we present the basic definitions of the theory of pseudo-Boolean optimization that are needed to reduce the objective function of a QSAP. These definitions are taken from [8]. We start by introducing the concept of pseudo-Boolean functions.

**Definition 5.27 (Pseudo-Boolean function)** *For $\mathbb{B} = \{0, 1\}$ and a fixed number $n \in \mathbb{N}$, a pseudo-Boolean function*

$$f : \mathbb{B}^n \to \mathbb{R}.$$

*maps a set of Boolean variables to the real numbers.*

To represent a pseudo-Boolean function, we introduce the concept of multi-linear polynomials. Note that each pseudo-Boolean function can be uniquely expressed as a multi-linear polynomial.

**Definition 5.28 (Multi-linear polynomial, Degree)** *For a given set $N = \{1, ..., n\}$, a multi-linear polynomial is of the form*

$$f(x_1, ..., x_n) = \sum_{S \subseteq N} c_S \prod_{j \in S} x_j$$

*with $x_i \in \mathbb{B}$ for all $i \in \{1, .., n\}$ and $c_S \in \mathbb{R}$ for all $S \subseteq N$. For this sum, we define the product over the empty set by*

$$\prod_{j \in \emptyset} x_j = 1.$$

*For a given multi-linear polynomial representation of a pseudo-Boolean function $f$, the degree of $f$ is defined as the size of the largest subset $S \subseteq N$ with $c_S \neq 0$,*

$$deg(f) = max\{d \in \mathbb{N}_0 \mid \exists S \subseteq N : |S| = d \land c_S \neq 0\}.$$

*A pseudo-Boolean function $f$ is called linear (quadratic, cubic, ...), if*

$$deg(f) \leq 1 \ (\leq 2, \leq 3, ...)$$

*holds.*

A different way to represent pseudo-Boolean functions are posiforms, i.e., polynomial expressions in terms of all the literals $x_i$ and $\overline{x}_i$ ($\overline{x}_i = (1 - x_i)$).

**Definition 5.29 (Posiform, Degree)** *Given the variables* $x_1, ..., x_n \in \mathbb{B}$
*and the set of literals* $L = \{x_1, \overline{x}_1, ..., x_n, \overline{x}_n\}$, *a posiform is defined by*

$$\phi(x_1, ..., x_n) = \sum_{T \subseteq L} a_T \prod_{u \in T} u$$

*with* $a_T \geq 0$ *if* $T \neq \emptyset$. *A posiform with* $a_T = 0$ *for* $T = \emptyset$ *is said to be
homogeneous. Similar to the degree of a multi-linear polynomial, the degree
of a posiform is defined by*

$$deg(\phi) = max\{d \in \mathbb{N}_0 \mid \exists T \subseteq L : |T| = d \wedge a_T \neq 0\}.$$

*We call a posiform* $\phi$ *linear (quadratic, cubic, ...), if*

$$deg(\phi) \leq 1 \ (\leq 2, \leq 3, ...)$$

holds.

It is customary to define $a_T = 0$ if $u, \overline{u} \in T$ holds for a literal $u \in L$, since
$x_i \in \mathbb{B}$ implies $u \cdot \overline{u} = 0$.

**Remark 5.30** *For the relation between posiforms and pseudo-Boolean functions it can be easily seen that every posiform uniquely defines a pseudo-
Boolean function. Conversely, a pseudo-Boolean function can have several
posiforms representing it.*

Hammer et al. introduced in [25] the concept of a height of a quadratic
pseudo-Boolean function $f$. It is defined as the greatest constant $c$ so that
there exists a homogeneous quadratic posiform $\phi(x, \overline{x})$ $(x = (x_1, ..., x_n))$ with

$$f(x) = c + \phi(x, \overline{x}) \quad \forall x \in \{0, 1\}^n.$$

We extend this definition by introducing different heights for the different
degrees of the allowed posiforms $\phi$. The concept of these new heights is
inspired by the different levels of the RLT formulation.

**Definition 5.31 (Height)** *For a given pseudo-Boolean function* $f$ *of degree
$t$, the height of degree $t$, denoted by $H_t(f)$, is defined as the greatest constant
$c$ so that there exists a homogeneous posiform $\phi(x, \overline{x})$ of degree $deg(\phi) = t$
with*

$$f(x) = c + \phi(x, \overline{x}) \quad \forall x \in \{0, 1\}^n.$$

A more specific and problem related version of the height is defined by $H_{t,X}$ for a set $X \subseteq \{0,1\}^n$. For a pseudo-Boolean function $f$ and a set $X \subseteq \{0,1\}^n$, the height $H_{t,X}[f(x)]$ is defined as the greatest constant $c$ so that there exists a homogeneous posiform $\phi(x, \overline{x})$ of degree $deg(\phi) = t$ for which

$$f(x) = c + \phi(x, \overline{x}) \quad \forall x \in X$$

holds.

For these two different versions of heights, the statements of the following proposition are straightforward to see.

**Proposition 5.32** *For a given pseudo-Boolean function $f(x)$ and a set $X \subseteq \{0,1\}^n$, the following inequalities hold:*

$$H_{t,X}[f(x)] \geq H_{t,\{0,1\}^n}[f(x)] = H_t(f) \quad \forall t \geq deg(f)$$
$$H_{t,X}[f(x)] \geq H_{t-1,X}[f(x)] \quad \forall t \geq deg(f) + 1,$$

*For a QSAP with the objective function $g(x)$, we get in addition*

$$H_{t,\mathcal{A}}[g(x)] \leq \min QSAP \quad \forall t \geq 2.$$

**Proof:**

The first inequality is clear, since $X \subseteq \{0,1\}^n$ holds. The second inequality follows directly from the fact that each multi-linear polynomial of degree $t-1$ is also of degree $t$. Finally, the third inequality is clear, due to the definition of the height.

$\square$

## 5.8.2  Reduction of QSAPs

With the definition of heights for pseudo-Boolean functions, we have an approach to extract lower bounds from the objective function of a QSAP. But in addition, we want to keep an eye on the change of the objective function while extracting a constant from it. For this, we introduce the concept of a reduction of an objective function. We only consider QSAPs in this section. Remark that a feasible solution of a QSAP is of the form

$$x_{ij} \in \{0,1\}, \quad \sum_{j=1}^{n_i} x_{ij} = 1 \quad \forall i \in M.$$

**Definition 5.33 (Reduction, Best Reduction)** *Given an objective function $f$ of a QSAP with the set of feasible solutions $\mathcal{A}$, a quadratic reduction $f'$ is a new objective function with nonnegative coefficients $b'_{ij}$ and $c'_{ijkl}$ for which*

$$f(x) = a' + f'(x)$$
$$f'(x) = \sum_{i=1}^{m} \sum_{j=1}^{n_i} b'_{ij} x_{ij} + \sum_{i=1}^{m-1} \sum_{k=i+1}^{m} \sum_{j=1}^{n_i} \sum_{l=1}^{n_k} c'_{ijkl} x_{ij} x_{kl}$$

*holds for all feasible assignments $x \in \mathcal{A}$.*

*A reduction is called best quadratic reduction, if the corresponding constant $a'$ is maximal, i.e., if there does not exist a constant $a^* > a'$ and a quadratic pseudo-Boolean function $f^*$ with nonnegative coefficients for which*

$$f(x) = a^* + f^*(x)$$

*holds for all feasible assignments $x \in \mathcal{A}$.*

*In general, a reduction is called quadratic (cubic, ...) if the reduced function $f'$ may have the degree 2 (3,...).*

For quadratic reductions and the quadratic pseudo-Boolean function

$$f(x) = q_0 + \sum_{i=1}^{n} q_i x_i + \sum_{i=1}^{n-1} \sum_{j=i+1}^{n} q_{ij} x_i x_j,$$

Hammer et al. proved in [25] that $H_2(f)$ is equal to the optimal solution value of the following linear program:

$$\min q_0 + \sum_{i=1}^{n} q_i x_i + \sum_{i=1}^{n-1} \sum_{j=i+1}^{n} q_{ij} y_{ij},$$
$$s.t. \ y_{ij} \geq 0 \quad \forall i,j \in \{1,...,n\} \ (i < j),$$
$$y_{ij} \leq x_i, \ y_{ij} \leq x_j \quad \forall i,j \in \{1,...,n\} \ (i < j),$$
$$1 - x_i - x_j + y_{ij} \geq 0 \quad \forall i,j \in \{1,...,n\} \ (i < j).$$

Billionnet and Elloumi extended this result in [5] so that, for a QSAP with the objective function $f(x)$, the height $H_{2,\mathcal{A}}[f(x)]$ is equal to the optimal solution of the corresponding RLT-1 formulation. Hence,

$$H_{2,\mathcal{A}}[f(x)] = \min \ \text{RLT-1}$$

holds for QSAPs.

We already know that the heights of different degrees are always feasible lower bounds of a QSAP. An interesting question is, whether there is a degree for which the height equals the optimal solution value of the QSAP. The following proposition gives an answer to this question.

**Proposition 5.34** *For a QSAP of size* $|M| = m$, *a feasible set of assignments* $\mathcal{A}$ *and an objective function* $f(x)$, *the equation*

$$H_{m,\mathcal{A}}[f(x)] = \min \ QSAP$$

*holds.*

**Proof:**

Since we regard the height of degree $m$, coefficients $a_T$ with $|T| = m$ that belong to $m$ literals are allowed in the reduction $f'$. Thus, we have a unique coefficient $d_{1j_1,\ldots,mj_m}$ for every solution $\sigma$ of the QSAP. By setting these coefficients to the solution value of the corresponding QSAP solution and by giving all other coefficients the value zero, we generate a reduction of the objective function.

We can extract the lowest of these coefficients $d_{i_1j_1,\ldots,i_mj_m}$ as a constant summand. Thus, we get a new reduction, since we do not change the total value for any solution $x \in \mathcal{A}$. It is easy to see that this extracted constant equals the optimal solution value of the QSAP. Thus, it is the maximal value that can be extracted, because of the inequality from Proposition 5.32. Hence, the extracted value is the height $H_{m,\mathcal{A}}$ of this QSAP, which proves the proposition.

□

We know from the previous proposition that the lowest degree $t'$ for which the height $H_{t',\mathcal{A}}[f(x)]$ is a tight lower bound for the QSAP must be between 2 and $m$, which is similar to the stepwise improving RLT formulations. The exact level for which the height returns tight lower bounds for every QSAP of a certain size is unknown, in contrast to the RLT theory, where the level $t$ for which the RLT-$t$ formulation is tight is known (cf. Theorem 5.16 of this thesis).

### 5.8.3 Applying the Reduction Technique to $G_{RLT}^t$

A first comparison between the reduction theory and the theory of the RLT can be made by analyzing the effect of different height levels on the problem graphs $G_{RLT}^t$ from Definition 5.11. These graphs can be used to construct QSAPs for which the RLT-$t$ formulation returns untight lower bounds.

We start with the heights of different levels for the QSAP that is generated by the graph $G_{RLT}^1$. This graph is presented by the non-dashed lines in Figure 5.15. The corresponding QSAP that we introduced in the proof of Lemma 5.14 has the objective function

$$f(x) = \sum_{i=1}^{m} \sum_{j=1}^{n_i} b_{ij} x_{ij} + \sum_{i=1}^{m-1} \sum_{k=i+1}^{m} \sum_{j=1}^{n_i} \sum_{l=1}^{n_k} c_{ijkl} x_{ij} x_{kl}$$

with coefficients

$$b_{ij} = 0,$$

$$c_{ijkl} = \begin{cases} 1 & \text{, if } e_{ijkl} \in E_{RLT}^1 \\ 0 & \text{, else.} \end{cases}$$

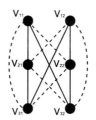

Figure 5.15: Visualization of $G_{RLT}^1$. The coefficients of the objective function of the QSAP that correspond to the edges of this graph (shown as non-dashed lines) get the assigned value $c_{ijkl} = 1$. Every solution of the QSAP contains at least one dashed line.

For the degree $t = 2$, we get the height

$$H_{2,\mathcal{A}}[f(x)] = 0$$

We can calculate the height $H_{3,\mathcal{A}}[f(x)]$ of degree $t = 3$ by introducing cubic coefficients in the objective functions of the form

$$d'_{112131} = 1, \quad d'_{112132} = 1, \quad d'_{112231} = 3, \quad d'_{112232} = 1,$$
$$d'_{122131} = 1, \quad d'_{122132} = 3, \quad d'_{122231} = 1, \quad d'_{122232} = 1.$$

Thus, we get the cubic reduction

$$f(x) = \sum_{j=1}^{2} \sum_{l=1}^{2} \sum_{q=3}^{2} d_{1j2l3q} x_{ij} x_{kl} x_{pq} \quad \forall x \in \mathcal{A}$$

and the cubic height of this function is equal to the smallest of these $d'$ coefficients. Hence, we have

$$H_{3,\mathcal{A}}[f(x)] = 1 = \min \text{QSAP}.$$

In general, we can see that the height of degree $(t+2)$ is a tight lower bound for the QSAP that results from the problem graph $G^t_{RLT}$. Thus, we have for these QSAPs the equation

$$H_{(t+2),\mathcal{A}}[f(x)] = \min \text{RLT-}(t+1) = \min \text{QSAP}.$$

## 5.8.4 A Dual Ascent Strategy for Generating Lower Bounds

This subsection presents a Dual Ascent approach to generate lower bounds for the QSAP. The approach is inspired by the Dual Ascent procedure of Adams et al. [1] and Hahn et al. [23].

The main idea is to increase the extracted constant value of the objective function stepwise by reorganizing the structure of the objective function. Therefore, we try to find good quadratic reductions. Additionally, even better bounds can be achieved by allowing reductions of higher levels. In our Dual Ascent approach, we allow cubic reductions and we try to approximate $H_{3,A}[f(x)]$ for an objective function $f(x)$ and a set of allowed assignments $\mathcal{A}$ of a QSAP. Since we use coefficients that correspond to variables of up to 3-cliques, the approach corresponds to the RLT-2 formulation. Thus, we also speak of a Dual Ascent level 2 algorithm.

We now describe in detail how the reformulation of the objective function is done in our approach. In addition to the concept of having coefficients for vertices and edges, we also allow coefficients $d'_{ijklpq}$ for vertex triples $t_{ijklpq} = \{v_{ij}, v_{kl}, v_{pq}\}$. At the start of the Dual Ascent algorithm, all these vertex triple coefficients are initialized with value zero, since we regard quadratic problems.

We start by redistributing the objective function coefficients of the vertices evenly over the coefficients of the adjacent edges so that all new vertex coefficients have value zero. This concept was already introduced in Proposition 3.9. Note that we only change the coefficients of the objective function without damaging its global structure. Thus, the new objective function is a reduction of the initial objective function $f(x)$.

The pseudocode of this step is presented in Figure 5.16. Note that, in all pseudocodes, we do not differ between the order of the index pairs. This means, e.g., that $c_{ijkl}$ refers to the same coefficient as $c_{klij}$. This is also the case for the higher coefficients.

---

**Algorithm 5.8.1:** UPWARDREDISTRIBUTIONOFVERTICES()

**for each** vertex $v_{ij}$ with $b'_{ij} > 0$

$\begin{cases} w_{red} \leftarrow \lfloor b'_{ij}/(m-1) \rfloor \\ w_{rest} \leftarrow b'_{ij} \mod (m-1) \\ \textbf{for } k \leftarrow 1 \textbf{ to } m \quad (k \neq i) \\ \quad \begin{cases} \textbf{for } l \leftarrow 1 \textbf{ to } n_k \\ \quad \begin{cases} c'_{ijkl} \leftarrow c'_{ijkl} + w_{red} \\ \textbf{if } (k == \max\{\{1, ..., m\} \setminus \{i\}\}) \\ \quad \textbf{then } \left\{ c'_{ijkl} \leftarrow c'_{ijkl} + w_{rest} \right. \end{cases} \\ b'_{ij} \leftarrow 0 \end{cases}$

---

Figure 5.16: Upward redistribution of the vertex coefficients to the incident edge coefficients.

After this redistribution of the values of the vertex coefficients to the edges, we now go one step further and redistribute the coefficient values from the edges to the vertex triples without changing the general objective function structure. This redistribution results in an objective function, in which all edge coefficients have value zero. The corresponding pseudocode of this step is presented in Figure 5.17.

---

**Algorithm 5.8.2:** UPWARDREDISTRIBUTIONOFEDGES()

**for each** edge $e_{ijkl}$ with $c'_{ijkl} > 0$

$\begin{cases} w_{red} \leftarrow \lfloor c'_{ijkl}/(m-2) \rfloor \\ w_{rest} \leftarrow c'_{ijkl} \mod (m-2) \\ \textbf{for } p \leftarrow 1 \textbf{ to } m \quad (p \neq i,\ p \neq k) \\ \quad \begin{cases} \textbf{for } q \leftarrow 1 \textbf{ to } n_p \\ \quad \begin{cases} d'_{ijklpq} \leftarrow d'_{ijklpq} + w_{red} \\ \textbf{if } (p == \max\{\{1,...,m\} \setminus \{i,k\}\}) \\ \quad \textbf{then } \left\{ d'_{ijklpq} \leftarrow d'_{ijklpq} + w_{rest} \right. \end{cases} \\ c'_{ijkl} \leftarrow 0 \end{cases} \end{cases}$

---

Figure 5.17: Upward redistribution of edge coefficients to adjacent triangle coefficients.

After applying these two redistribution steps to the QSAP, we get a new objective function that consists only of cubical coefficients $d'_{ijklpq}$ and an extracted constant value $a'$ (after the first upward redistribution, $a'$ still has value zero). The next step of the Dual Ascent algorithm is to redistribute the objective function coefficients downwards back to the edge and vertex coefficients. For this, we start by ordering the edges and vertices. This can be done randomly to avoid getting stuck in a local minimum or by a heuristic. Afterwards, each of the edges (vertices) extracts the highest possible value from its adjacent vertex triples (edges) so that the structure of the objective function is not changed. Finally, the highest possible constant value is extracted from the vertex coefficients $b'_{ij}$ and this value is added to the lower bound value $a'$. This process is presented in the pseudocode in Figure 5.18. Note that the vertex triple coefficients do not necessarily have value zero after this downward redistribution step. Thus, the corresponding Semi-Assignment Problem has a cubic objective function now.

In the Dual Ascent algorithm, the upward and downward redistribution steps are done alternately. Thus, the lower bound is increased stepwise until a stopping criterion is fulfilled. Such a criterion is in our case that no im-

---

**Algorithm 5.8.3:** DOWNWARDREDISTRIBUTION()

Order all edges $E = \{e_1, e_2, ...\}$.
**for each** $e_{ijkl} \in E$

$\left\{ \begin{array}{l} \textbf{for } p \leftarrow 1 \textbf{ to } m \quad (p \neq i, \; p \neq k) \\ \left\{ \begin{array}{l} d_{\min} \leftarrow \min_{q \in N_p} d'_{ijklpq} \\ c'_{ijkl} \leftarrow c'_{ijkl} + d_{\min} \\ \textbf{for } q \leftarrow 1 \textbf{ to } n_p \\ \left\{ d'_{ijklpq} \leftarrow d'_{ijklpq} - d_{\min} \right. \end{array} \right. \end{array} \right.$

Order all vertices $V = \{v_1, v_2, ...\}$.
**for each** $v_{ij} \in V$

$\left\{ \begin{array}{l} \textbf{for } k \leftarrow 1 \textbf{ to } m \quad (k \neq i) \\ \left\{ \begin{array}{l} c_{\min} \leftarrow \min_{l \in N_k} c'_{ijkl} \\ b'_{ij} \leftarrow b'_{ij} + c_{\min} \\ \textbf{for } l \leftarrow 1 \textbf{ to } n_k \\ \left\{ c'_{ijkl} \leftarrow c'_{ijkl} - c_{\min} \right. \end{array} \right. \end{array} \right.$

**for** $i \leftarrow 1$ **to** $m$

$\left\{ \begin{array}{l} b_{\min} \leftarrow \min_{j \in N_i} b'_{ij} \\ lowerBound \leftarrow lowerBound + b_{\min} \\ \textbf{for } j \leftarrow 1 \textbf{ to } n_i \\ \left\{ b'_{ij} \leftarrow b'_{ij} - b_{\min} \right. \end{array} \right.$

---

Figure 5.18: Downward redistribution from the vertex triple coefficients to the lower bound.

provement of the lower bound can be achieved in several consecutive steps. The flowchart of this Dual Ascent algorithm is presented in Figure 5.19.

Several tests of this algorithm on test instances show promising results. A good convergence behavior can be seen, especially in the beginning of the runtime. The computational results of the Dual Ascent level 2 algorithm are presented in Section 6.3.5.

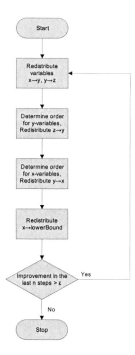

Figure 5.19: Flowchart of the RLT Dual Ascent level 2 algorithm.

## 5.8.5   Dual Ascent 2.5

In Section 5.7, we introduced an approach that creates an RLT formulation that combines the simplicity of the RLT-1 formulation with the tightness of the RLT-2 formulation. With the presented RLT-1.5 algorithm, a good compromise between the strengths of both formulations could be achieved.

For the Dual Ascent algorithm, a similar problem occurs. For bigger problem sizes, the Dual Ascent level 2 algorithm that was presented in Section 5.8.4 is not able to generate sufficient lower bounds for the given test instances. But generating a Dual Ascent algorithm that works on a higher level (for example the Dual Ascent level 3 algorithm that includes all possible 4-cliques) results in a model, whose size is too large to be manageable.

Therefore, a good compromise between the level 2 and the level 3 Dual Ascent approach is needed. Such a trade-off is presented in this subsection, where we introduce the Dual Ascent 2.5 algorithm. We begin with the standard Dual Ascent level 2 formulation and then introduce stepwise parts of the level 3 formulation so that the algorithm is able to extract better lower bounds in the following steps.

The algorithm starts with the Dual Ascent level 2 formulation. But if the stopping criterion (no improvement in the last few steps) is reached, the algorithm does not stop. Instead, an upward redistribution is made and the four rows with the highest average row triple coefficients between them are determined (if the four rows were already chosen in an earlier improvement step, we choose the next best four rows that are not used for an extension step yet). For these four rows, all coefficients that correspond to 4-cliques are generated and introduced into the model. Here, we generate new coefficients $e'_{ijklpqrs}$ with an initial value of zero. We choose the four rows with the highest average row triple coefficients, since they refer to a situation where a high potential for further good extractions is given.

While doing an upward redistribution, all coefficients of vertex triples whose vertices are from these four rows get an extra upward redistribution step as it is shown in Figure 5.20. Here, the value of a vertex triple coefficient is evenly redistributed to the coefficients of the corresponding vertex quadruple coefficients. After this step, such a vertex triple coefficient also has the value zero.

With these new coefficients in the objective function, we also need a different downward redistribution process. Thus, before making a downward redistribution for the edges, we have to make a downward redistribution for all vertex triples $t_{ijklpq} = \{v_{ij}, v_{kl}, v_{pq}\}$ that are a subset of a vertex quadruple that corresponds to one of the generated $e'_{ijklpqrs}$ coefficients. The pseudocode of this additional downward redistribution is presented in Figure 5.21. If the approach reaches again the first stopping criterion (no improvements in the last few steps), then another row quadruple is determined and further coefficients are introduced in the objective function. We repeat this process until a second stopping criterion is fulfilled (e.g., no improvements are achieved within the last $n'$ extensions of the objective function with further coefficients).

The flowchart of the Dual Ascent level 2.5 algorithm is presented in Figure 5.22. Several tests of the approach on test instances show that the Dual Ascent level 2.5 algorithm further improves the Dual Ascent level 2 algorithm.

---

**Algorithm 5.8.4:** UPWARDREDISTRIBUTIONOFTRIPLES()

**for each** vertex triple $t_{ijklpq}$ with $d'_{ijklpq} > 0$ that belongs to a set of introduced quadruples ($\nu_{i,k,p}$ is the number of quadruples in which the rows $i$, $k$ and $p$ are included)

$\begin{cases} w_{red} \leftarrow \lfloor d'_{ijklpq}/\nu_{i,k,p} \rfloor \\ w_{rest} \leftarrow d'_{ijklpq} \mod (\nu_{i,k,p}) \\ \textbf{for } r \leftarrow 1 \textbf{ to } m \quad \text{with } (i,k,p,r) \text{ is a used row quadruple} \\ \quad \begin{cases} \textbf{for } s \leftarrow 1 \textbf{ to } n_r \\ \quad \begin{cases} e'_{ijklpqrs} \leftarrow e'_{ijklpqrs} + w_{red} \\ \textbf{if } (r == \max\{\{1,...,m\} \setminus \{i,k,p\}\}) \\ \quad \textbf{then } \left\{ e'_{ijklpqrs} \leftarrow e'_{ijklpqrs} + w_{rest} \right. \end{cases} \end{cases} \\ d'_{ijklpq} \leftarrow 0 \end{cases}$

---

Figure 5.20: Upward redistribution of vertex triple coefficients to the corresponding vertex quadruple coefficients.

---

**Algorithm 5.8.5:** DOWNWARDREDISTRIBUTION()

Order vertex triples that belong to introduced row quadruple.
**for each** $t_{ijklpq} \in orderedTriples$
$\begin{cases} \textbf{for } r \leftarrow 1 \textbf{ to } m \quad (\{i,k,p,r\} \text{ is a used row quadruple}) \\ \quad \begin{cases} e_{\min} \leftarrow \min_{s \in N_r} e'_{ijklpqrs} \\ d'_{ijklpq} \leftarrow d'_{ijklpq} + e_{\min} \\ \textbf{for } s \leftarrow 1 \textbf{ to } n_r \\ \quad \left\{ e'_{ijklpqrs} \leftarrow e'_{ijklpqrs} - e_{\min} \right. \end{cases} \end{cases}$

---

Figure 5.21: Downward redistribution from the quadruple coefficients to the triangle coefficients.

Detailed computational results of the Dual Ascent level 2.5 approach and a comparison to other lower bound approaches for the QSAP are presented in

Section 6.3.6.

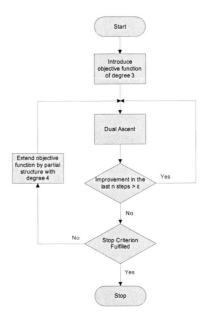

Figure 5.22: RLT Dual Ascent 2.5 Algorithm.

There is still room for further improvement of the algorithm. The order in which we make the downward redistribution is generated randomly at the moment. By a further analysis of the structure of the coefficients, there may exist further room for improvement. Another part of the algorithm that can be optimized is the choice of the row quadruples for which we introduce new coefficients. Here, an analysis that is similar to the one for the RLT-1.5 algorithm may help to find better combinations that further improve our approach.

## 5.8.6 Hybrid Dual Ascent - RLT Approach

A different approach that can be used to enhance the bounds of the Dual Ascent level 2 algorithm is to hybridize the approach with the RLT-1.5 formulation. Here, we take the quadratic part

$$g'(x) = a' + \sum_{i=1}^{m} \sum_{j=1}^{n_i} b'_{ij} x_{ij} + \sum_{i=1}^{m-1} \sum_{k=i+1}^{m} \sum_{j=1}^{n_i} \sum_{l=1}^{n_k} c'_{ijkl} x_{ij} x_{kl}$$

of the reduction of the objective function that is generated by the Dual
Ascent approach. We use this function $g'(x)$ as the objective function for the
RLT-1.5 approach. The cubic coefficients of the objective function $g'(x)$ are
disregarded.

Nevertheless, the hybrid algorithm is able to generate better improvements of the Dual Ascent level 2 algorithm than the Dual Ascent level 2.5
approach. In general, the best achieved bounds for the large test instances are
gained with this hybrid approach. The detailed results and the performance
of our hybrid algorithm are presented in Section 6.3.7.

## 5.9   Summary of the Chapter

In this chapter, we introduced the concept of the Reformulation Linearization
Technique to generate good lower bounds for the QSAP. In addition to the
RLT-1 formulation and the RLT-2 formulation, we present a new notation
for the abstract RLT-$t$ formulation.

The RLT provides a stepwise tightening formulation for the QSAP, but
the smallest level $t$ for which the RLT-$t$ formulation is tight for a certain
QSAP size was up to now unknown. We give an answer to this question
by proving the minimal level $\gamma(m, N)$ that fulfills this tightness requirement.
This is done by analyzing the RLT as a problem on graphs. Here, we present
for each RLT level $t$ a minimal problem graph $G_{RLT}^t$ that is responsible for
an untight formulation.

The gained knowledge of this concept allows us to generate an improved
RLT-1.5 algorithm that stepwise eliminates these problem structures from
the RLT-1 formulation by successively inserting RLT-2 constraints into the
model.

Another concept to generate good bounds for the QSAP is to approximate
the height $H_{t,A}[f(x)]$ of the objective function of a QSAP. The approach is
based on the theory of pseudo-Boolean functions. It also provides a stepwise
improving series of lower bounds, like the RLT formulation. To approximate

the height, we present a Dual Ascent algorithm that successively extracts constant values from the objective function.

To improve the Dual Ascent approach, we use the idea that was used for the RLT-1.5 algorithm to combine the simplicity of the lower level formulation with the tightness of the higher level formulation. Here, we present a Dual Ascent level 2.5 algorithm that starts by approximating the height $H_{3,\mathcal{A}}[f(x)]$ of degree three. If the improvements of the algorithm get too low, we stepwise allow more changes to the objective function to get an approximation of $H_{4,\mathcal{A}}[f(x)]$.

Finally, we present a hybrid approach that combines the Dual Ascent level 2 approach and the RLT-1.5 formulation. Here, we use the transformed function of the Dual Ascent algorithm as an objective function for the RLT formulation.

The presented approaches show promising results if used as lower bound techniques for the manually generated test instances and for the real life test instances.

# Chapter 6

# Computational Results

In this chapter, we analyze the efficiency of the previously introduced solution strategies for the QSAP. The algorithms are tested on a set of problem instances, not necessarily resulting from TTSPs. The focus of the analysis concentrates on solution quality and convergence speed. The objective function that we use for the real life test instances is the percentage improvement of $\psi_1$ that represents the overall solution quality of a timetable.

The computations of the test runs are executed on a Microsoft Windows Server 2003 with 15.9 GB of memory and an Intel Xeon E5420 processor with 2.5 GHz. All implementations are written in C# and are compiled as 32 bit executables with a Microsoft Visual Studio 2005 compiler. The mixed integer programs and linear programs are solved with Ilog Cplex 11.2.

## 6.1   Problem Instances

We begin by introducing the QSAP test instances that we use in this thesis to analyze the quality of the different approaches. Most of these instances are randomly generated and are not based on real life timetable data. The instances differ in their size and in the structure of the objective function. The problem instances have sizes between 5 and 31 objects and between 1 and 20 possible locations for each object.

Most of the test instances are manually created. Here, we do not use coefficients $b_{ij}$ in the objective function, since we can redistribute them on the edge coefficients $c_{ijkl}$ as it was shown in Proposition 3.9. Thus, all objective functions are of the form

$$f(x) = \sum_{i=1}^{m-1} \sum_{k=i+1}^{m} \sum_{j=1}^{n_i} \sum_{l=1}^{n_k} c_{ijkl} x_{ij} x_{kl}.$$

We distinguish between two types of structures for the objective functions of the test instances. The first structure type has random coefficients $c_{ijkl}$ between a minimal value $c_{\min}$ and a maximal value $c_{\max}$. But if we regard real life TTSPs, the resulting QSAPs do not have such a random structure. Thus, the second type of objective function has a structure, in which the coefficients $c_{ijkl}$ for two objects (this means for fixed $i$ and $k$) have similar values. To get such a structure, we generate a balanced objective function with coefficients $c_{ijkl}$ that may differ from an average value $\hat{c}_{ik}$ only by a maximum percentage $p$. In this allowed interval, the values are uniformly distributed. The values $\hat{c}_{ik}$ of the manually created test instances are chosen randomly between a minimal value $\underline{c}$ and a maximal value $\bar{c}$.

We present the list of the test instances that we use in Table 6.1. For each problem size, there is a balanced and an unbalanced test instance. The values $c_{ijkl}$ of the balanced test instances may differ from the average values $\hat{c}_{ik}$ by 20%.

| Problem Instance | Objects | Locations | Properties |
|---|---|---|---|
| QSAP_m5n5 | 5 | 5 | $\underline{c} = 50,\ \bar{c} = 100,\ p = 20\%$ |
| QSAP_m15n5 | 15 | 5 | $\underline{c} = 50,\ \bar{c} = 100,\ p = 20\%$ |
| QSAP_m10n10 | 10 | 10 | $\underline{c} = 50,\ \bar{c} = 100,\ p = 20\%$ |
| QSAP_m15n10 | 15 | 10 | $\underline{c} = 50,\ \bar{c} = 100,\ p = 20\%$ |
| QSAP_m15n15 | 15 | 15 | $\underline{c} = 50,\ \bar{c} = 100,\ p = 20\%$ |
| QSAP_m5n5+ | 5 | 5 | $c_{\min} = 0,\ c_{\max} = 100$ |
| QSAP_m15n5+ | 15 | 5 | $c_{\min} = 0,\ c_{\max} = 100$ |
| QSAP_m10n10+ | 10 | 10 | $c_{\min} = 0,\ c_{\max} = 100$ |
| QSAP_m15n10+ | 15 | 10 | $c_{\min} = 0,\ c_{\max} = 100$ |
| QSAP_m15n15+ | 15 | 15 | $c_{\min} = 0,\ c_{\max} = 100$ |

Table 6.1: Test instances for the QSAP. The values $c_{ijkl}$ are uniformly distributed between $c_{\min}$ and $c_{\max}$ if the properties contain no value for $p$.

We also consider real life problems of different sizes. The smallest example refers to the city Pirmasens ($\approx 40,000$ inhabitants) with only 9 lines. Furthermore, we use two test instances for the city of Kaiserslautern ($\approx 100,000$

inhabitants) with 14 lines and 31 lines. Note that the number of possible shifts is not the same for all lines, since, e.g., we do not allow the train lines to be shifted. Thus, we have different values $n_i$ in the QSAPs that result from real life TTSPs.

| Problem Instance | Lines | Possible shifts |
|---|---|---|
| Pirmasens | 9 | 1-15 |
| Kaiserslautern medium (KL_med) | 14 | 1-15 |
| Kaiserslautern big (KL_big) | 31 | 1-15 |

Table 6.2: Real life test instances for the QSAP.

## 6.2 Exact Solution Strategies

As an exact solution strategy, we use the Mixed Integer Program (MIP) formulation from Section 3.4. This approach generates the optimal solution values that we compare later with the results of the other approaches that generate lower bounds for the QSAP.

### 6.2.1 MIP Formulation

Solving the MIP formulation generates an optimal solution of a QSAP. Hence, it is desirable to use this approach to solve the QSAPs that result from the TTSP. Unfortunately, the QSAP is NP-hard. Therefore, exact solution strategies are not a good choice for bigger problem sizes, since the computational requirements are too high. Nevertheless, the results of the MIP formulation can help to analyze the quality of the heuristics and the relaxed formulations.

Since the MIP solver that we use applies mainly a branch and cut approach, we also get lower bounds during the runtime of the calculation. Thus, we can compare the lower bounds of the MIP optimization with the lower bounds of the relaxed formulations at different points in time to get an estimation of the convergence speed of the algorithms. Note that the MIP-solvers are generally not created to efficiently find good lower bounds. This should be considered while comparing the lower bound estimations.

The results and the computation times of the MIP approach in comparison to the relaxed formulations are presented in Table 6.3 (for the manually generated test instances) and in Table 6.4 (for the real life test instances). It can be seen that the computation times grow exponentially with the problem size. For bigger problem sizes, the computation of the MIP solution takes more than one day. For the largest unbalanced problem, the solver was not able to find the optimal solution at all.

We present some graphics to compare the speed of convergence of the MIP and the other approaches. In Figure 6.2, the developments of the lower bounds of the MIP and of the RLT-1.5 algorithm are compared for the KL_big real life test instance. It can be seen that the RLT-1.5 approach returns better bounds in the first 10,000 seconds.

The progress of the estimated lower bounds in the first few minutes of the runtime of the algorithms is very important, since we need good lower bounds quickly. The first 200 seconds of the calculation of the lower bounds from the MIP formulation, the Dual Ascent level 2 algorithm and the RLT-1.5 approach for the KL_big test instance are compared in Figure 6.4. The Cplex solver uses a precomputation for an efficient calculation of the MIP solution; this explains that the first results occur here after approximately 100 seconds.

In Section 6.5, some results of the ACO metaheuristic are compared to the optimal MIP solutions. In Figure 6.6, we present the approximated Pareto frontier of the multi-colony ACO with an MIP bound for $\psi_1$. Here, the optimal solution is found by the metaheuristic. In the Figures 6.7 and 6.8, we compare the approximated Pareto frontier that is generated by the metaheuristic with optimal MIP solutions for linear combinations of the objective functions. Thus, we can compare the approximated Pareto frontier in multiple dimensions with the lower bounds of the MIP formulation.

## 6.3   Lower Bound Approaches

Unlike exact solution strategies, the approaches in this section generate lower bounds for the QSAP. Here, not only the quality of the generated bounds is important, but also the time that is needed to generate these bounds. Additionally, the progress of the algorithms during the runtime is interesting. This refers to the ability of improving the lower bounds mainly in the first phase of the algorithm and refining them at the end, in contrast to a linear enhancement of the lower bounds over time.

We analyze the results of the following approaches:

- the Trivial Bounds from Section 3.10,

- the RLT-1 formulation from Section 5.4.1 (which is equivalent to the Linear Program (LP) formulation from Section 3.4.2),

- the RLT-2 formulation from Section 5.4.2,

- the RLT-1.5 approach from Section 5.7,

- the Dual Ascent level 2 algorithm from Section 5.8.4,

- the Dual Ascent level 2.5 algorithm from Section 5.8.5,

- the hybrid Dual Ascent level 2 and RLT-1.5 algorithm from Section 5.8.6.

The bounds for the test instances, the computation times of these bounds and the progress that the algorithms make during the computation are presented in Section 6.4.

## 6.3.1 Trivial Bounds

Trivial bounds are a good approach to get a first estimation of the lower bounds of a QSAP. The computation of the results can be done quickly in a few seconds, even for the big test instances. A disadvantage of these bounds is that the gaps of the solutions are too large for most problems to be of a good use in, e.g., branch and bound algorithms.

We analyze the four different trivial bounds Row-Pair Bound, Triangle Bound, Star Bound and Pair-Star Bound from Section 3.10. The Pair-Star Bound generates the best results of these four bounds, but the computational efforts of this bound type are also the highest.

The results of the trivial bounds are presented in Table 6.7 for the manually generated test instances and in Table 6.8 for the real life test instances. It can be seen that the bound quality differs a lot for the test instances. Especially for the unbalanced QSAP instances, the trivial bounds perform prohibitively bad. Even the pair-star bound has gaps of more than 50%. Nevertheless, for balanced test instances like the real life test instances, the trivial bounds can be used to get a first estimation of the solution quality. Here, the gaps of the pair-star bound are less than 5%.

## 6.3.2   RLT-1 Formulation

The RLT-1 formulation returns in most cases better lower bounds than the trivial bounds. Nevertheless, the gaps of these lower bounds can still be significantly high. Note that the size of the gaps is problem dependent. It can be seen that the gaps are much higher for unbalanced problems.

The calculation times for the RLT-1 approach are very small compared to the exact MIP approach, but they are still much higher than the times that the trivial bounds need. The results of the RLT-1 formulation are similar to the ones of the pair-star bound for the manually generated test instances. But for the real life test instances, the RLT-1 formulation performs better, resulting in a gap of less than 3.5% for the KL_big problem instance.

The detailed results of the RLT-1 formulation can be found in Table 6.3 (for the manually generated test instances) and in Table 6.4 (for the real life test instances).

## 6.3.3   RLT-2 Formulation

The RLT-2 formulation is a stronger linear formulation of a QSAP than the RLT-1 formulation. It is able to generate good lower bounds, but the problem size is very high for bigger QSAPs. For most of our test instances, the RLT-2 approach is able to calculate exact bounds, but the computation times of the LP solvers are even longer than the ones of the MIP solvers. This is the case, because of the high number of constraints and variables that are needed for the RLT-2 formulation. The computation time for the QSAP_m10n10 instance is much bigger than the one for the QSAP_m15n5 instance, since the first instance has $10^{10}$ possible solutions and the second one has only $15^5$ solutions ($10^{10} \approx 15^5 \cdot 1.3 \cdot 10^5$).

In contrast to the RLT-1 formulation, there is no big difference between the balanced and the unbalanced test instances. Only for the QSAP_m15n5 and the QSAP_m15n5+ instances, the generated lower bounds are not equal to the optimal solution value of the MIP. For bigger problem sizes, it is not possible to generate the RLT-2 solution, because of the high number of variables. Even for the medium sized $QSAP\_m10n10$ and $QSAP\_m10n10+$ test instances, the solver needs several hours to solve the RLT-2 formulation. The detailed results of the RLT-2 formulation for the test instances are presented in Table 6.3 and in Table 6.4.

Figure 6.1 shows the progress of the RLT-2 bounds during the runtime of the algorithm for the KL_med instance. It can be seen that the lower bounds of the RLT-2 formulation increase in an approximately linear way in the first 8000 seconds after the initial progress in the first 100 seconds. Here, the RLT-1.5 approach has a much better behavior. This can be seen if we compare the two curves that show the enhancement of the lower bounds during the runtime for both approaches.

### 6.3.4 RLT-1.5 Formulation

The RLT-1.5 approach combines the simplicity of the RLT-1 formulation with the good solution quality of the RLT-2 formulation. Thus, it has the ability to generate good lower bounds in a short time. Afterwards, it refines these bounds by introducing new RLT-2 constraints into the model.

The results of the RLT-1.5 algorithm are presented in Table 6.5 for the manually generated test instances and in Table 6.6 for the real life test instances. The tables also show the improvements that can be achieved in comparison to the RLT-1 formulation. Especially for the unbalanced and the medium sized real life problems, significant improvements of the lower bounds can be achieved.

In Figure 6.1, we compare the improvement rate of the RLT-1.5 algorithm and the RLT-2 formulation for the KL_big instance. It can be seen that the RLT-1.5 approach performs significantly better than the RLT-2 approach.

The RLT-1.5 algorithm returns better bounds during the early phase of the runtime compared to the MIP approach. This is presented in Figure 6.2. Here, the bounds of the RLT-1.5 approach are better in the first 10,000 seconds.

The convergence behavior of the algorithm can be seen in Figure 6.3 for the KL_big instance. Here, the points in time where new constraints are considered are easily identifiable, since the improvement rate of the algorithm is much faster after new constraints are introduced into the model.

We present the enhancements of the bounds during the runtime of the Dual Ascent level 2 algorithm, the RLT-1.5 approach and the MIP formulation for the real life test instance KL_big in Figure 6.4. Here, the Dual Ascent approach performs best, since it benefits most from the balanced structure of the problem.

### 6.3.5   Dual Ascent Level 2 Formulation

The Dual Ascent level 2 algorithm (DA2) is an approach that generates good lower bounds in a short time by reformulating the objective function. It extracts a constant value from the objective function which we can use as a lower bound for the QSAP. The algorithm contains random decisions. Therefore, the outcome of a single run of the algorithm cannot be predicted. Nevertheless, the results of the algorithm are quite stable. This can be seen in Table 6.9 and in Table 6.10, where the outcome of ten runs for each of the test instances is presented. Here, we see that the generated lower bounds are better than the ones of the RLT-1 formulation or the trivial bounds. In addition to the good bound quality, the computational efforts that are needed to compute these bounds is acceptable. For the KL_med real life test instance, the algorithm is able to compute the optimal solution value. This implies that, in this case, the height $H_{3,\mathcal{A}}[(f(x)]$ of degree three corresponds to the optimal solution value of the QSAP.

A comparison of the development of the bounds during the runtime of the different algorithms is presented in Figure 6.4 for the KL_big real life test instance. Our DA2 approach shows a promising behavior and outperforms the RLT-1.5 formulation and the MIP approach for this test instance.

A comparison of the DA2 algorithm, the extended Dual Ascent level 2.5 approach and the hybrid DA2 and RLT-1.5 approach is presented in Figure 6.5. Here, the extended level 2.5 version and the hybrid approach are able to improve the bound quality of the DA2 approach significantly.

### 6.3.6   Dual Ascent Level 2.5 Formulation

The Dual Ascent level 2.5 formulation is an enhancement of the level 2 formulation. Similar to the RLT-1.5 approach, it improves the formulation by introducing stepwise elements of the formulation of the next level. Like the Dual Ascent level 2 approach, the level 2.5 formulation is also nondeterministic.

The improvement rates of this approach are presented in Table 6.11 for the manually generated test instances and in Table 6.12 for the real life test instances. Here, the enhancements of the gaps of the Dual Ascent level 2 approach are presented. The gaps can be improved by up to 1.8%, depending on the problem structure.

The progress of the approach compared to the Dual Ascent level 2 approach and the hybrid approach is presented in Figure 6.5. Here, the hybrid approach performs best, but the Dual Ascent level 2.5 algorithm is also able to improve the bound quality significantly for the analyzed big real life test instance.

### 6.3.7 Hybdrid Dual Ascent Level 2 and RLT-1.5 Approach

Another technique to improve the generated bounds of the Dual Ascent level 2 algorithm is to use a hybrid approach that contains the RLT-1.5 algorithm. The improvements that can be achieved with this approach are presented in Table 6.11 and in Table 6.12. The enhancements of the gaps dominate the Dual Ascent level 2.5 approach. Improvements of more than 4% can be achieved.

The efficiency of the approach is presented in Figure 6.5, where the progress of the bounds can be seen over the runtime of the algorithm. Here, the hybrid approach is able to improve the bound quality of the Dual Ascent level 2 approach much faster than the Dual Ascent level 2.5 approach.

These results show that the hybrid algorithm is the most promising of the presented approaches. It combines the fast utilization of the problem structure by the Dual Ascent approach with the analytical strength of the RLT-1.5 approach.

## 6.4 Comparison of the Results

In this section, we analyze and compare the results of the different approaches. This is done with respect to two main aspects. The first one is the quality of the solutions and the computation times for the different formulations. The second aspect is the convergence speed of the different approaches during their runtime.

### 6.4.1 Solution Quality

We analyze now the quality of the best found solutions and the computed lower bounds of the different approaches. Additionally, we have a look at the computation times that are needed to compute these bounds.

The computation times and the solution gaps of the generated lower bounds for the MIP formulation, the RLT-1 formulation and the RLT-2 formulation are presented in Table 6.3 for the manually generated test instances. It can be seen that the RLT-1 formulation shows significant differences if used for the balanced and the unbalanced problems. For the unbalanced test instances, we have gaps up to 56%. The RLT-2 formulation is able to generate better solutions, due to its tighter formulation. But the computation time of the RLT-2 formulation grows exponentially with the problem size. Thus, this approach cannot be used to solve larger problem instances.

| Problem Instance | MIP | RLT-1 | | RLT-2 | |
|---|---|---|---|---|---|
| | Time | Gap | Time | Gap | Time |
| QSAP_m5n5 | 0.48 | 1.98% | 0.03 | 0% | 0.30 |
| QSAP_m15n5 | 44.94 | 5.08% | 0.50 | 0.36% | 4600.00 |
| QSAP_m10n10 | 45.11 | 4.95% | 0.75 | 0% | 16,707.11 |
| QSAP_m15n10 | 18,080.60 | 6.90% | 3.60 | - | - |
| QSAP_m15n15 | >86,400.00 | 7.49% | 18.33 | - | - |
| QSAP_m5n5+ | 0.05 | 16.29% | 0.05 | 0% | 0.11 |
| QSAP_m15n5+ | 38.84 | 33,73% | 0.73 | 1.87% | 3,879.74 |
| QSAP_m10n10+ | 51.45 | 46.15% | 0.89 | 0% | 19,753.94 |
| QSAP_m15n10+ | 20,778.86 | 55.87% | 2.58 | - | - |
| QSAP_m15n15+ | >86,400.00 | -* | 9.59 | - | - |

Table 6.3: Computation times and gaps of the lower bounds for the different formulations. The times are presented in seconds. Computations of the RLT-2 approach are aborted after 24 hours (86,400 seconds).
*The optimal solution of the QSAP_m15n15+ test instance is unknown. Thus, the gap of the RLT-1 bound is also unknown.

As expected, the results for the real life test instances are similar to the results for the balanced QSAP instances. This can be seen in Table 6.4. The RLT-2 formulation is able to find the optimal solution values of the QSAPs that correspond to the Pirmasens and the KL_med test instances. For the small Pirmasens example and the medium-sized KL_med example, the computation times of the MIP formulation are acceptable if only one objective function is considered. Nevertheless, seven seconds is a very long time if the optimization must be run several times to approximate the three-dimensional Pareto frontier in the objective space via weighted sums of the objective functions.

| Problem Instance | MIP | RLT-1 | | RLT-2 | |
|---|---|---|---|---|---|
| | Time | Gap | Time | Gap | Time |
| Pirmasens | 0.67 | 0.37% | 0.47 | 0% | 27.91 |
| KL_med | 7.78 | 3.57% | 1.4 | 0% | 38,580.00 |
| KL_big | >86,400.00 | 3.23% | 39.44 | - | - |

Table 6.4: Optimal solutions, lower bounds and gaps of the different approaches for the real life test instances. The times are presented in seconds.

The RLT-1.5 approach is a compromise between the RLT-1 formulation and the RLT-2 formulation. Its strength comes from solving several improving linear program formulations. The results for the given test instances are presented in Table 6.5 for the manually generated test instances and in Table 6.6 for the real life test instances. We let the RLT-1.5 algorithm run ten times as long as the linear program solver needs for the RLT-1 formulation for the different test instances. The results show that significant improvements to the results of the RLT-1 formulation can be achieved, especially for the unbalanced test instances. Note that the RLT-1.5 approach is able to generate even better bounds if the runtime is increased.

A different approach for generating lower bounds are the trivial bounds from Section 3.10. The gaps of these trivial bounds for the manually generated test instances are presented in Table 6.7. The Pair Star bound generates, as expected, the best results. The results show that the trivial bounds perform much better for the balanced problems. Here, gaps of less than 7% can be achieved. This result is similar to the one of the RLT-1 approach.

The results for the real life test instances are presented in Table 6.8. Promising results can be found in less than a second, even for the large KL_big instance. Here, the balanced structure is responsible for the good results for real life QSAPs.

Lower bounds with gaps that are smaller than five percent are quite good compared to the small computational costs of the trivial bounds. But for the TTSP, gaps of five percent are a significant difference. Remark that the TTSP already starts with a good synchronization and improvements of up

| Problem Instance | Gap | Diff. to RLT-1 Gap |
|---|---|---|
| QSAP_m5n5 | 0.0% | -1.98% |
| QSAP_m15n5 | 3.67% | -1.41% |
| QSAP_m10n10 | 3.53% | -1.42% |
| QSAP_m15n10 | 6.19% | -0.71% |
| QSAP_m15n15 | 7.01% | -0.48% |
| QSAP_m5n5+ | 0.0% | -16.29% |
| QSAP_m15n5+ | 20.17% | -13.56% |
| QSAP_m10n10+ | 35.26% | -10.89% |
| QSAP_m15n10+ | 50.49% | -5.38% |

Table 6.5: Gaps of the lower bounds gained from the RLT-1.5 algorithm for the manually generated test instances. We compare the results to the bounds gained by the RLT-1 formulation. The RLT-1.5 algorithm is allowed to run ten times as long as the solver needs for the RLT-1 formulation.

| Problem Instance | Gap | Diff. to RLT-1 Gap |
|---|---|---|
| Pirmasens | 0.0% | -0.37% |
| KL_med | 0.59% | -2.98% |
| KL_big | 2.96% | -0.27% |

Table 6.6: Gaps of the lower bounds gained from the RLT-1.5 algorithm for the real life test instances. We compare the results to the bounds gained by the RLT-1 formulation. The RLT-1.5 algorithm is allowed to run ten times as long as the solver needs for the RLT-1 formulation.

| Problem Instance | Pair Bound | Triangle | Star | Pair Star |
|---|---|---|---|---|
| QSAP_m5n5 | 7.98% | 4.98% | 5.94% | 2.89% |
| QSAP_m15n5 | 12.83% | 9.60% | 9.17% | 5.77% |
| QSAP_m10n10 | 10.17% | 8.46% | 8.64% | 5.29% |
| QSAP_m15n10 | 12.08% | 10.21% | 9.99% | 6.76% |
| QSAP_m15n15 | 11.00% | 9.98% | 9.96% | 6.83% |
| QSAP_m5n5+ | 86.84% | 54.76% | 56.20% | 23.79% |
| QSAP_m15n5+ | 90.78% | 66.29% | 65.41% | 39.13% |
| QSAP_m10n10+ | 97.91% | 77.96% | 79.62% | 46.63% |
| QSAP_m15n10+ | 98.12% | 82.31% | 79.42% | 54.13% |

Table 6.7: Gaps of the trivial lower bounds for the manually generated test instances.

| Problem | Pair Bound | Triangle | Star | Pair Star |
|---|---|---|---|---|
| Pirmasens | 8.09% | 5.42% | 5.68% | 3.50% |
| KL_med | 8.39% | 6.58% | 6.25% | 4.45% |
| KL_big | 7.10% | 5.80% | 5.39% | 4.20% |

Table 6.8: Gaps of the trivial lower bounds for the real life test instances.

to 6% can be achieved (cf. Figure 6.6). Here, lower bound gaps of 5% are too much to estimate an approximated Pareto frontier properly.

Better bounds than the trivial bounds or the bounds gained by the RLT-1 formulation can be generated with the Dual Ascent level 2 approach. In Table 6.9 (for the manually generated test instances) and in Table 6.10 (for the real life test instances), we present the best achieved gap, the worst achieved gap and the average gap of the lower bounds generated in ten runs for each of the different test instances.

| Problem Instance | min gap | average gap | max gap | average time |
|---|---|---|---|---|
| QSAP_m5n5 | 0.34% | 0.90% | 1.36% | 0.20 |
| QSAP_m15n5 | 4.18% | 4.36% | 4.45% | 2.24 |
| QSAP_m10n10 | 3.02% | 3.23% | 3.39% | 8.42 |
| QSAP_m15n10 | 6.02% | 6.10% | 6.19% | 35.39 |
| QSAP_m15n15 | 6.83% | 7.02% | 7.24% | 122.38 |
| QSAP_m5n5+ | 0.75% | 2.26% | 4.89% | 0.32 |
| QSAP_m15n5+ | 13.91% | 14.31% | 14.59% | 4.89 |
| QSAP_m10n10+ | 16.29% | 17.76% | 19.65% | 9.68 |
| QSAP_m15n10+ | 29.50% | 30.10% | 30.46% | 55.72 |

Table 6.9: Gaps and computation times of the Dual Ascent level 2 algorithm for the manually generated test instances. Each problem instance is optimized ten times and we present the minimal, the maximal and the average gaps.

| Problem Instance | min gap | average gap | max gap | average time |
|---|---|---|---|---|
| Pirmasens | 0.01% | 0.05% | 0.08% | 2.37 |
| KL_med | 0.0% | 0.001% | 0.004% | 18.55 |
| KL_big | 2.17% | 2.19% | 2.22% | 739.86 |

Table 6.10: Gaps and computation times of the Dual Ascent level 2 algorithm for the real life test instances. Each problem instance is optimized ten times and we present the minimal, the maximal and the average gaps.

It can be seen that the algorithm shows a very stable behavior. The difference between the minimal and the maximal found lower bounds is in most cases small. If we compare the results with the bounds of the RLT-1 formulation and the trivial bounds, we see a significant improvement. Especially for the unbalanced and the real life test instances, the results are promising. The algorithm is also able to generate the optimal solution value as a lower bound for the KL_med test instance.

For the Dual Ascent level 2.5 algorithm and the hybrid approach that both improve the results of the Dual Ascent level 2 algorithm, we present the achieved average improvements in Table 6.11 for the manually generated test instances and in Table 6.12 for the real life test instances. The improvements of the lower bounds after one hour of runtime are taken into account for several test runs. It can be seen that the hybrid approach dominates the Dual Ascent level 2.5 approach. This is the case, since the modern solvers that are used in the RLT-1.5 approach can utilize the problem structure better than the nondeterministic redistributions of the Dual Ascent approach. Nevertheless, the Dual Ascent approach nearly reached the same improvement for the important big real life test instance.

The improvements of the bounds are larger for the unbalanced test instances. This is plausible, since there is more room for improvement here, due to the higher gaps of the Dual Ascent level 2 bounds. The hybrid approach is able to generate an improvement of more than 4 % for the unbalanced QSAP_m10n10+ test instance.

| Problem Instance | DA2.5 | Hybrid |
|---|---|---|
| QSAP_m5n5 | -0.42% | -0.85% |
| QSAP_m15n5 | -0.24% | -1.00% |
| QSAP_m10n10 | -0.43% | -1.90% |
| QSAP_m15n10 | -0.12% | -0.83% |
| QSAP_m15n15 | -0.00% | -0.58% |
| QSAP_m5n5+ | -1.28% | -1.69% |
| QSAP_m15n5+ | -1.82% | -3.03% |
| QSAP_m10n10+ | -1.23% | -4.10% |
| QSAP_m15n10+ | -1.05% | -2.20% |

Table 6.11: Improvements of the gaps of the Dual Ascent level 2 approach gained by the Dual Ascent level 2.5 approach and the hybrid Dual Ascent and RLT-1.5 algorithm for the manually generated test instances. The average improvements for several runs are taken into account.

| Problem Instance | DA2.5 | Hybrid |
|------------------|-------|--------|
| Pirmasens | -0.02% | -0.03% |
| KL_med | -0.00% | -0.00% |
| KL_big | -0.31% | -0.35% |

Table 6.12: Improvements of the gaps of the Dual Ascent level 2 approach gained by the Dual Ascent level 2.5 approach and the hybrid Dual Ascent and RLT-1.5 approach for the real life test instances. The average improvements for several runs are taken into account. Note that the KL_med test instance is already solved to optimality by the Dual Ascent level 2 algorithm.

## 6.4.2   Speed of Convergence Analysis

For the analysis of the algorithms' progress during their runtime, we present several figures that visualize the speed of convergence of the different approaches. Therefore, we compare the lower bound values of the approaches at several points in time for different problem instances. All graphics in this section use the runtime in seconds as $x$-axis. The $y$-axis shows the percentage of the generated lower bounds from the optimal solution value.

Figure 6.1 compares the convergence behavior of the RLT-1.5 approach and the RLT-2 approach for the KL_med problem instance. It can be seen that the RLT-1.5 algorithm converges quite fast (to a gap of less than 0.75%). The curve of the RLT-2 algorithm has a nearly linear structure in the first 8,000 seconds after the initial progress in the first 100 seconds.

The behavior of the RLT-1.5 approach in comparison to the lower bounds computed by the MIP formulation for the KL_big instance is presented in Figure 6.2.  The RLT-1.5 approach returns better bounds than the MIP formulation during the first 9,000 seconds. This emphasizes that the RLT-1.5 algorithm is a good choice for generating lower bounds for the QSAP, since it strengthens the LP formulation and it returns better bounds than the MIP formulation in the early phase of the computation.

In Figure 6.3, we visualize the stepwise progress of the RLT-1.5 algorithm. The approach consists of a consecutive optimization of stepwise tightening formulations. When one of these formulations is solved, the algorithm enhances the formulation by additional constraints and variables. These points in time can be easily identified, since there is a more substantial improvement of the lower bounds after such a reformulation step.

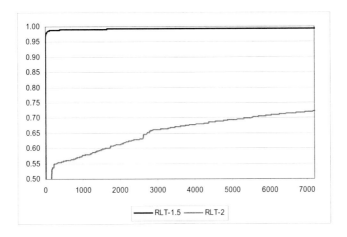

Figure 6.1: Comparison of the RLT-2 approach and the RLT-1.5 approach for the KL_med real life test instance. The $y$-axis shows the lower bound of the overall situation $\psi_1$ divided by the optimal solution value.

Up to now, we analyzed the development of the lower bounds only for longer time periods. Figure 6.4 shows the behavior of the Dual Ascent level 2 approach, the RLT-1.5 algorithm and the MIP formulation for the KL_big problem instance during the first 200 seconds. Here, the strength of the DA algorithm to generate good bounds quickly is evident. The RLT-1.5 approach starts similar to the DA approach, but the progress decreases faster. The MIP has a precomputation phase of 100 seconds for the large problem instance. After this precomputation, it starts with a bound arising from the LP relaxation and it has only a weak increase.

To compare the Dual Ascent level 2 approach and the Dual Ascent 2.5 approach, we let both algorithms run for the KL_big example. To analyze the improvement that is achieved by the additional redistribution steps, we disable the stopping criterion of the level 2 approach. Such a comparison for the KL_big test instance is shown in Figure 6.5. The convergence behavior in the first phase is similar for all three approaches, since they all start with the simple Dual Ascent level 2 approach. But when the level 2 algorithms stagnates, the Dual Ascent level 2.5 approach and the hybrid approach are both able to improve the lower bound significantly. Here, the improvements

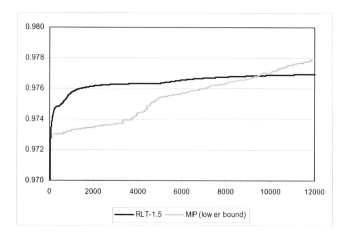

Figure 6.2: Comparison of the RLT-1.5 approach and the lower bounds generated by the solver of the MIP approach for the KL_big instance. The $y$-axis shows the lower bound of the overall situation $\psi_1$ divided by the optimal solution value.

of the hybrid approach are gained much faster than the ones of the Dual Ascent level 2.5 approach.

## 6.5   Metaheuristics and Lower Bounds

In this section, we analyze the quality of the solutions that are generated by the hybridized multi-colony ACO metaheuristic from Section 4.3 by comparing them to the solutions of the MIP formulation. The closer the solutions are to the optimal lower bounds that are generated by the MIP formulation, the better the approximation of the Pareto frontier is for this objective function. In Figure 6.6, such an MIP bound is shown for the KL_big example. The ACO approach finds the solution that corresponds to this optimal bound after a computation time of less than one hour. This is quite fast, MIP solvers need several days to find this solution.

Nevertheless, the result that the optimal solution for $\psi_1$ is reached does not necessarily mean that the approximated Pareto frontier of the meta-

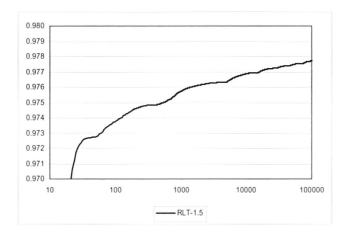

Figure 6.3: Convergence behavior of the RLT-1.5 algorithm for the KL_big test instance. The $x$-axis shows the computation time in seconds on a logarithmic scale. The $y$-axis shows the lower bound of the overall situation $\psi_1$ divided by the optimal solution value. The points in time where new RLT-2 constraints are introduced into the model can be identified by the higher improvement rate that results from the tighter formulations.

heuristic has in general a good structure. Thus, we also try to find bounds for the multiobjective approach.

For the following analysis, we have to modify the line variance from Section 2.5, since this function cannot be transformed into a binary formulation of the form

$$f(x) = \sum_{i=1}^{m-1} \sum_{k=i+1}^{m} \sum_{j=1}^{n_i} \sum_{l=1}^{n_k} c_{ijkl} x_{ij} x_{kl}.$$

Instead, we regard in this section a binary line variance that is defined by the sum of the pairwise shift differences within the different line clusters. This function has a binary formulation $f(x)$ with

$$c_{ijkl} = \begin{cases} |j - l| & \text{, if line } i \text{ and line } k \text{ are in the same line cluster,} \\ 0 & \text{, else.} \end{cases}$$

We assume that all lines of a line cluster have the same shift interval. If this is not the case, we can replace $|j - l|$ by $|s_{ij} - s_{kl}|$.

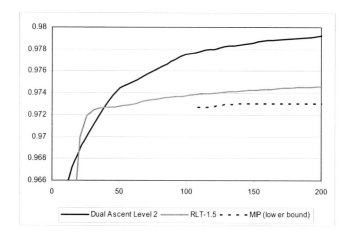

Figure 6.4: Comparison of the Dual Ascent level 2 approach, the RLT-1.5 approach and the lower bounds of the MIP formulation for the KL_big instance at the early phase of the optimization. The $x$-axis shows the runtime of the approaches in seconds, the $y$-axis shows the lower bound of the overall situation $\psi_1$ divided by the optimal solution value. Note that the MIP has a precomputation phase. Hence, it generates the first bounds after 100 seconds.

Given this new line variance, we can calculate multidimensional lower bounds. For this, we approximate the Pareto frontier from the lower bound side via hyperplanes that are given by linear combinations of the objective functions. Such a lower bound approach is presented in Figure 6.7. The left graphic shows the set of the best solutions that are found by the ACO metaheuristic for the KL_small instance. The right figure shows the generated lower bound hyperplanes. No solutions can exist bottom left of these hyperplanes. These bounds are generated by the MIP formulation. Thus, there must exist at least one Pareto optimal solution on each of these hyperplanes. It can be seen that the solutions of the metaheuristic get very close to the Pareto frontier, which indicates a good quality of the approximated Pareto frontier.

For bigger problem sizes, it is computationally expensive to generate these hyperplanes. For the KL_big example, each run of the MIP solver runs for a

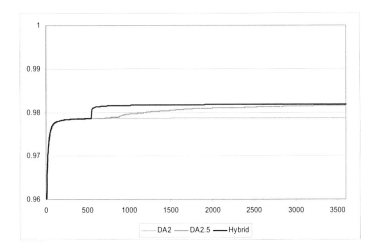

Figure 6.5: Visualization of the Dual Ascent level 2 approach, the Dual Ascent level 2.5 algorithm and the hybrid approach. The $x$-axis shows the runtime of the approaches in seconds, the $y$-axis shows the lower bound of the overall situation $\psi_1$ divided by the optimal solution value.

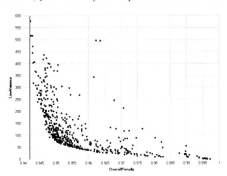

Figure 6.6: Visualization of the non-dominated solutions generated by the ACO metaheuristic for the KL_big example after one hour. The $x$-axis shows the percentage improvements of $\psi_1$, the $y$-axis presents $\psi_3$. The black vertical line shows the optimal solution value for $\psi_1$.

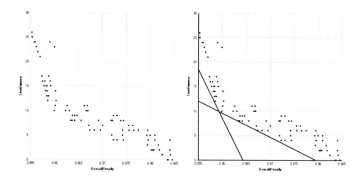

Figure 6.7: Visualization of the approximated Pareto frontier of the ACO metaheuristic for the KL_small example. The $x$-axis shows $\psi_1$, the $y$-axis presents the binary line variance. The bold lines in the right figure show the optimal lower bound hyperplanes generated by the MIP formulation.

single hyperplane takes more than two days of computation time. Hence, this approach is not applicable for a practical usage. Nevertheless, the results are interesting for us to measure the quality of a metaheuristic. In Figure 6.8, an approximated Pareto frontier that is calculated by the hybridized ACO metaheuristic is presented for the large real life KL_big test instance. In the right figure, the lower bound hyperplanes are added and we can see that the solution quality of the metaheuristic is promising, even for large real life TTSPs.

But for practical purposes we have to estimate faster, whether an approximated Pareto frontier has a good quality. To solve this problem, we compare the calculated solutions of a metaheuristic with hyperplanes that are generated by one of the presented lower bound approaches. For the example that we show in Figure 6.9, we take the hybrid Dual Ascent level 2 and RLT-1.5 algorithm. Here, the bounds that we can calculate in much less time than the MIP bounds also give a good estimation of the exact position of the Pareto frontier. Thus, they allow a quick evaluation of the solution quality.

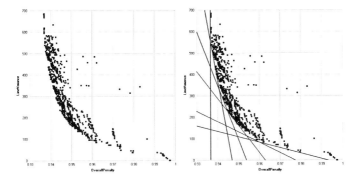

Figure 6.8: Visualization of the approximated Pareto frontier of the ACO metaheuristic for the KL_big test instance. The $x$-axis shows the percentage improvement of $\psi_1$ to the current timetable, the $y$-axis presents the binary line variance. The bold lines in the right figure show the optimal lower bounds generated by the MIP formulation.

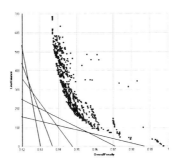

Figure 6.9: Visualization of the approximated Pareto frontier of the ACO metaheuristic for the KL_big example. The $x$-axis shows $\psi_1$, the $y$-axis presents the binary line variance. The bold lines present the lower bounds that are generated by the hybrid Dual Ascent level 2 and RLT-1.5 approach.

## 6.6 Summary of the Chapter

In this chapter, we apply the different lower bound approaches to various QSAP test instances. Here, we differ between balanced and unbalanced problem structures. Most of the test instances are manually generated, but we also test the approaches on real life data for TTSPs of different sizes.

The approaches perform significantly better for the balanced test instances. Since the real life TTSP has also a balanced structure, we are able to achieve good lower bounds here as well. The computation times of the different approaches differ a lot. Thus, one has to find a good trade-off between the computation time and the bound quality, when choosing a lower bound approach for the TTSP. Here, the hybrid Dual Ascent level 2 and RLT-1.5 approach performs best.

In addition to the analysis of the bound quality and the computation time, we also have a look at the progress that the different approaches make during their runtime. This helps to estimate the efficiency of the algorithms.

We analyze the quality of the approximated Pareto frontier of the hybrid multi-colony ACO metaheuristic from Chapter 4 with multidimensional lower bound approaches. The results show that the metaheuristic is able to find Pareto optimal solutions for real life TTSPs in an acceptable time.

# Chapter 7

# A Real Life Application

The mathematical concepts that are presented in this work are part of the software solution "SynPlan". Here, traffic planners get assistance based on mathematical techniques that help them to synchronize the timetables efficiently.

In this chapter, the main ideas of the software concept and an application to the urban area of the city of Kaiserslautern are presented. The results of the optimization process are analyzed with respect to their practicability. In addition, we compare the two different objective function approaches from Section 2.5.

## 7.1 The Public Transport Network of Kaiserslautern

Kaiserslautern is a medium-sized city in south western Germany with about 100,000 inhabitants. The public transport network is mainly organized by a traffic association ("Verkehrsverbund Rhein Neckar" (VRN)) that coordinates the different public transport companies. One task of the traffic association is to care for a good synchronization of the timetables at the intersection points of these companies.

Kaiserslautern is situated in the middle of the train line between the bigger cities Mannheim and Saarbruecken and has therefore a good transport connection to the railway system. Kaiserslautern is in addition the urban center of the region Westpfalz, to which commuters travel from the smaller suburbs by regional trains and regional buses.

The lines, network nodes and the waiting time classifications for this analysis are chosen in cooperation with the experts of the traffic association.

### 7.1.1   Traffic Companies

Most of the public transport in Kaiserslautern and the surrounding area is of one of the following three types: City buses, regional buses or regional trains. Different companies operate these modes of transportation, so a decentralized synchronization at the intersection points of the company networks is needed. In our analysis we consider

- 14 city bus lines (with mostly periodic timetables),

- 12 regional bus lines (with mostly aperiodic timetables),

- 9 regional train lines (with periodic and aperiodic timetables).

These three different transportation types are provided by different companies. These companies are regarded as the line clusters (lines that are bundled due to a similar structure) in our approach.

### 7.1.2   Network Nodes

The analysis includes three network nodes of different sizes. Each of these network nodes has a specific purpose in the network.

The main station that forms a network node with a nearby bus stop incorporates the train lines and connects them to the city bus network and the regional bus network.

The city center covers a wide area. Here, the passengers are willing to walk between five different stations to get good connections. At a certain station of this network node, the city buses are especially synchronized with several common meeting points. These common meeting points occur during the peak period on workingdays every 15 minutes. The city center is also the main starting point of the regional buses.

Finally, the western train station with a nearby bus stop is regarded in the analysis, since the traffic planners hope that there is room for improvement for the transfers between these two stations in the current timetable.

## 7.1.3   Waiting Time Evaluation

For all occurring triples (arriving line cluster, departing line cluster, network node), a waiting time classification must be made. For the chosen network nodes and line clusters, we have 22 waiting time classifications (nine at the main station, nine at the western train station and four at the city center, since the trains do not stop at this network node). Some examples of waiting time classifications are presented in Table 7.1.

| network node | connection type | almost | risk | conv. | patience |
|---|---|---|---|---|---|
| main station | train / city bus | -10 to -1 | 0 to 2 | 3 to 7 | 8 to 12 |
| city center | city bus / city bus | -5 to -1 | 0 to 1 | 2 to 3 | 4 to 12 |
| station west | reg. bus / train | -10 to -1 | 0 to 2 | 3 to 5 | 6 to 12 |
| ... | ... | ... | ... | ... | ... |

Table 7.1: Example of three of the 22 waiting time classifications. The time intervals (given in minutes) define the different transfer types.

As can be seen in the table, the risk intervals and the convenience intervals are sometimes chosen quite small. Especially the convenience interval for the city bus connection at the city center shows, how strict the transfers are planned in some timetables. In the current situation, the buses are planned quite efficiently and the bus drivers wait for each other. Thus, this small waiting time is sufficient in the majority of cases to give the passengers a convenient transfer.

For the different transfer types, we use the following penalty values as a basis for the penalty functions:

- almost transfer: 22

- risk transfer: 18

- convenience transfer: 1

- patience transfer: 10

- no transfer: 20

The values of the smoothed penalty function (cf. Figure 2.4) are modified by the factors that correspond to the ABC classification of the network node, the connection type and the connection.

### 7.1.4    Analyzed Time Intervals

The timetables vary a lot for the different time intervals. For example, the traffic during the rush-hour in the morning consists mainly of commuters that travel from the suburbs to the city. Conversely, most of the traffic during the rush-hour in the evening leads from the city back to the suburbs. These two situations need different focuses for the optimization. In addition, the situation on weekends is again totally different due to the absence of commuter traffic.

Therefore, we define several different time intervals for the analysis of the timetables. It would not make sense to include all of these intervals in one big analysis, since we cannot rank the connections properly in such a case. Additionally, the optimization goals and the importance of connections differ a lot for the different time intervals.

For our analysis of the city of Kaiserslautern, we define the following time intervals:

- workingday morning 4 a.m.– 10 a.m. (11.536 transfers)

- workingday midday 10 a.m.– 3 p.m. (10.874 transfers)

- workingday evening 3 p.m.– 8 p.m. (10.580 transfers)

- Saturday morning 5 a.m.– 12 p.m. (9.555 transfers)

- Saturday afternoon 12 p.m.– 8 p.m. (9.397 transfers)

- Sunday 8 a.m.– 6 p.m. (6.765 transfers)

The problem sizes of the different time intervals are presented by the number of transfers that these intervals contain.

### 7.1.5    ABC Classification

For each of these time intervals, an ABC classification for the three different levels network node, connection type and connection has been made. These classifications reflect the importance of the different connections in the network (e.g., number of passengers) and the focus of the analysis. To simplify the input of this data, we start with a general classification that assigns to all levels the importance 'B' and the traffic planner just has to mark the objects that are important ('A'), that are of minor importance ('C') or that should be disregarded ('I').

## 7.2 Software Concept "SynPlan"

The given approach from the previous chapters is implemented in the software concept "SynPlan". Here, an interactive optimization algorithm is combined with tools to analyze timetables for a given network.

This software combines mathematical algorithms, decision support for the traffic planner and intelligent ways to analyze both, the global and the local aspects of a timetable. In addition, the multidimensional character of the problem is integrated according to its high relevance.

### 7.2.1 Decision Support for the Timetable Synchronization Problem

It is important for an optimization software to present the solutions in a clear way. In the case of the TTSP, the traffic planner must be able to analyze both, the global and the local situation. For the global situation, the numbers of the different transfer types (risk transfer, convenience transfer, ...) and the overall convenience measure $\psi_1$ can help to get a good estimation of the solution quality. But for the local view, just showing numbers is not enough. Hence, we present the situation at the single connections in comprehensible graphics (cf. Figure 7.1). Here, all arriving tours and all departing tours of a connection are shown vertically arranged. Those tours that are connected by a transfer have a diagonal connector, whose gradient refers to the walking time between the stations. The color of the connector shows, which transfer type is present in the given situation. Small colored dots at the left end of the connector show in addition the situation of the unshifted timetable. Thus, the traffic planner can spot worsenings and improvements of the new situation easily.

There are a lot of decisions to make for the traffic planners while they synchronize timetables. The model must be adjusted and constraints must be set. But the most important decisions, which finally influence the real life timetables, are the shifts that are assigned to the individual lines. There is a high number of lines and for each line there are several possible shifts. Thus,

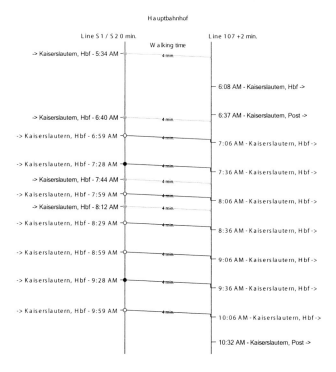

Figure 7.1: Visualization of a connection for the TTSP. The situation of the connection between the train line S1/S2 and the city bus line 107 at the main station for a morning of a workingday is shown. Line 107 is shifted by +2 minutes. The diagonal lines in the center of the figure represent transfers. The black lines are convenience transfers, the gray lines are no transfers. The small dots at the left side of these lines mark a change in the transfer quality. Here, the unfilled dots represent former risk transfers, the filled dots represent unchanged transfer qualities. Altogether, five risk transfers are improved to convenience transfers.

it is quite hard for the traffic planner to know which shift is promising. On a local level, one can easily improve single transfers manually, but automated decision support is needed to keep an eye on the global situation.

Here, a forecast of the consequences for the possible shifts of a line helps to avoid a frustrating trial and error search. The traffic planner can choose several aspects of the timetable that he wants to observe. These aspects can be the whole network, specific network nodes, connection types or single connections. For these chosen aspects of the timetable, the planner directly gets the information, which shifts have a promising outcome. Such a decision support graphic is presented in Figure 7.2.

Figure 7.2: Decision support for the shifts of a particular bus line. The different shifts are the small marks on the horizontal axis and the vertical dashed line represents the current shift situation. The black line refers to the changes of $\psi_1$, the gray line shows the changes of the number of convenience transfers. Higher values refer in both cases to a better situation.

These curves also have the positive side effect that they can additionally be used to analyze the computed solutions of the metaheuristics. The user can directly check, whether further improvements of the situation can be reached by small changes.

## 7.2.2 Constraint Handling

With the help of constraints, the traffic planner adds individual goals to the model. These constraints reduce the search space for the mathematical algorithms. The constraints are introduced in the software in an intuitive way, the traffic planner does not have to interact directly with the mathematical model. Instead, he directly places the constraints at the situation that he wants to preserve or where he wants to generate an improvement. This could be a train line that may not get a negative shift due to track limitations (lineshift constraint), a synchronization at an important network node that should be preserved (period constraint) or an almost transfer that should be changed so that it becomes a real transfer for the passengers (connection constraint).

These constraints are automatically integrated into the model. Thus, they change the behavior of the metaheuristics, even during their runtime. Such a change during the runtime causes the problem that some solutions are not feasible anymore. These solutions are removed and the metaheuristic proceeds its run based on the still feasible solutions.

This reuse of already found solutions has a clear advantage in comparison to a new start of the algorithm, since the information of the remaining solutions can be reused (e.g., as parents for the Genetic Algorithm or as a set of non-dominated ants that influence the pheromone trails for the Ant Colony Optimization approach).

## 7.3   Outcome of the Real Life Application

In this section, we present computational results for the different objective function approaches from Section 2.5. The first approach uses comfort related objective functions, meaning that they are based on an evaluation function that maximizes the average comfort of the passengers. The second approach uses transfer related objective functions, meaning that we compute the actual numbers of the generated transfer type changes.

We use the multi-colony ACO metaheuristic from Section 4.3 that is hybridized with a Local Search approach to generate the solutions that are presented in this section.

The graphics in the following subsection contain two-dimensional visualizations of approximated Pareto frontiers. Every point in such a graphic corresponds to a non-dominated solution. The solutions can be easily transformed into timetables for the underlying TTSP.

### 7.3.1   Comfort Related Objective Functions

The objective functions that we use in this subsection are the overall network quality $\psi_1$, the maximum percentage losses at a network node $\psi_2$ and the line variance $\psi_3$ that measures the amount of changes that are made to the current timetable. For $\psi_1$, we present the results in a slightly different form. For this, we divide the solution value by the value of $\psi_1$ for the current timetable situation. Thus, we get a percentage improvement which is more intuitive than only presenting a high number.

Even if the current timetables are already synchronized, there is still room for improvement. An optimization without constraints allows improvements to the overall situation of up to six percent. But even if the common meeting points in the city center are maintained, which is a high restriction to the search space, timetables with an overall improvement of 1.5% can still be found for the Kaiserslautern network. This is a high value if we consider the fact that the traffic planners invest great manual efforts to improve single transfers. Therefore, even small percentage improvements represent useful results for the traffic planners.

The objective functions $\psi_1$ and $\psi_3$ have a negative correlation. This means that for solutions with a better overall situation, more changes have to be made to the system. This correlation can be seen in Figure 7.3 (for the Kaiserslautern example with constraints that preserve the common meeting points at the city center) and in Figure 7.4 (the same example without constraints). Here, a two-dimensional projection of the three-dimensional approximated Pareto frontier for the workingday morning time interval is presented. In the first figure, we can see a clear structure of the frontier which disperses for the problem without constraints.

In Subsection 2.5.4, we mentioned the correlation of the objective functions $\psi_1$ and $\psi_2$. This correlation is illustrated in Figure 7.5 for the weekday morning time interval with a constraint that preserves the common meeting points at the city center and in Figure 7.6 for the example without constraints. The $x$-axis shows the percentage improvement of the overall situation and the $y$-axis presents $\psi_2$, the percentage gains or losses at the worst network node. The main diagonal shows the borderline, under which no solutions can exist, since the worst percentage situation at a network node cannot be better than the overall percentage situation.

It is remarkable that the solutions get quite close to the main diagonal, which refers to results that improve the timetable evenly at all network nodes. While regarding Figure 7.5, a differentiation between $\psi_1$ and $\psi_2$ seems dispensable, since both objective functions have similar values for most solutions. But in Figure 7.6, we can see that there can occur big differences in the values of these two objective functions for the generated solutions. The best solutions in terms of $\psi_1$ have a worsening for $\psi_2$ of 3% till 7%.

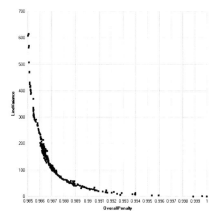

Figure 7.3: Visualization of the generated non-dominated solutions for the Kaiserslautern timetable (workingday morning) with a constraint that preserves the common meeting points of the city bus lines at the city center. The $x$-axis shows the percentage improvements of $\psi_1$, the $y$-axis presents $\psi_3$, the amount of changes that is needed to generate these new timetables.

## 7.3.2   Transfer Related Objective Functions

The results for the transfer related objective functions (cf. Section 2.5.4) are presented in this subsection. These objective functions are generated by choosing a level of detail (the overall situation, a network node, a connection type or a connection), a minimum importance ("A", "B" or "C") for the network nodes, the connection types and the connections and a set of regarded transfer type changes (old transfer type to new transfer type). Maintaining the status quo via a transfer type change 'convenience transfer to convenience transfer' is also possible. The algorithm tries to optimize (maximize if desirable changes are regarded or minimize if undesirable changes are regarded) the number of these chosen changes. Some examples of transfer related objective functions that we use are

- Maximize the number of gained convenience transfers with connection rank "A",

- Minimize the number of lost convenience transfers with connection rank

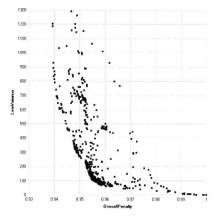

Figure 7.4: Visualization of the generated non-dominated solutions for the Kaiserslautern timetable (workingday morning) without constraints. The $x$-axis shows the percentage improvements of $\psi_1$, the $y$-axis presents $\psi_3$.

"A",

- Maximize the number of gained "real transfers",

- Minimize the number of lost "real transfers",

A gained convenience transfer is a convenience transfer that is not convenient in the current timetable. A "real transfer" is a transfer of the type risk, convenience or patience.

It is not possible to formulate the wish to improve the overall situation with a single objective function of this type. Thus, we combine these objective functions with $\psi_1$, the goal to optimize the overall network quality. Therefore, all optimization runs in this section are done with $\psi_1$ as the third objective function.

Figure 7.7 visualizes a two-dimensional representation of the approximated Pareto frontier with the total number of gained "real transfers" (a change from a "no transfer" or an "almost transfer" to a "risk transfer", a "convenience transfer" or a "patience transfer") as $x$-axis and the total number of lost "real transfers" as $y$-axis. The analysis is done for a workingday evening. It can be seen that all solutions are close to the main diagonal. Thus,

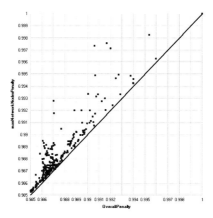

Figure 7.5: Visualization of the generated non-dominated solutions for the Kaiserslautern timetable (workingday morning) with a constraint that preserves the important bus synchronization at the city center. No solutions can exist below the diagonal borderline. The $x$-axis shows the percentage improvements of $\psi_1$, the $y$-axis presents $\psi_2$, the worst percentage situation at a single network node.

they have a similar amount of gained and lost transfers. This is caused by the interconnectedness of the network, where nearly all changes that create new "real transfers" also generate a loss of "real transfers" at other connections. Therefore, we have to differ between gained transfers that are really desirable and transfer losses that are acceptable to achieve these gains.

To get this differentiation, we regard in the following analysis the number of gained convenience transfers with connection rank "A" ($x$-axis) and the number of lost convenience transfers with connection rank "A" ($y$-axis). With this approach, we focus our analysis on the important improvements and worsenings of transfers. The outcome of the optimization run is presented in Figure 7.8. Here, most solutions are situated significantly under the main diagonal, meaning that there are more gained than lost important transfers. Most of the found solutions are located in a corridor that is parallel to the main diagonal. Since the traffic planner wants to preserve preferably many important transfers, especially the area marked by the ellipse in the bottom left corner is promising. Here, each loss of an important convenience transfer

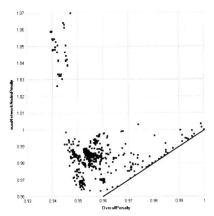

Figure 7.6: Visualization of the generated non-dominated solutions for the Kaiserslautern timetable (workingday morning) without constraints. The diagonal line shows the border under which no solutions can exist. The $x$-axis shows the percentage improvements of $\psi_1$, the $y$-axis presents $\psi_2$.

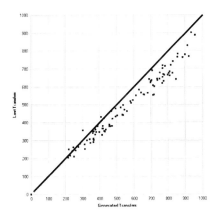

Figure 7.7: Optimization result for the workingday evening timetable. The graphic compares the number of generated "real transfers" ($x$-axis) versus the number of lost "real transfers" ($y$-axis).

is compensated by a gain of several new important transfers.

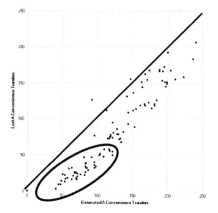

Figure 7.8: Optimization result for the workingday evening timetable. The graphic compares the number of generated convenience transfers with rank "A" ($x$-axis) and the number of lost convenience transfers with rank "A" ($y$-axis).

The same analysis is done for the Saturday morning period from 5 a.m. till 12 p.m. (noon). In Figure 7.9, we can find again the outcome that the number of gained "real transfers" is close to the number of lost "real transfers" for all solutions. In general, such a behavior can be found for all time intervals that we regard for the network of Kaiserslautern.

While analyzing the number of generated and lost "convenience" transfers with connection rank "A", a difference between the Saturday situation and the workingday situation can be found. Figure 7.10 shows the solutions of the Saturday morning analysis. Here, the solutions are much more scattered around the main diagonal. This is caused by the smaller number of tours on weekends, which makes it difficult to improve the situation locally without damaging the whole network. Nevertheless, we can also find a promising area in the bottom left corner of the figure (the area is again marked by an ellipse). Here, the ratio between generated transfers and lost transfers is promising.

In general, we can find such an area that contains a lot of feasible solutions with a good ratio between the number of gained important convenience

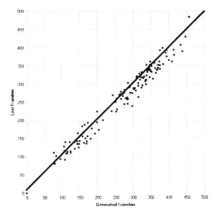

Figure 7.9: Optimization result for the Saturday morning timetable. The graphic compares the number of generated "real transfers" ($x$-axis) and the number of lost "real" transfers ($y$-axis).

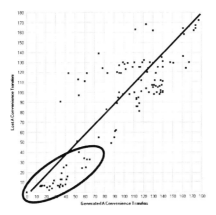

Figure 7.10: Optimization result for the Saturday morning timetable. The graphic compares the number of generated convenience transfers with rank "A" ($x$-axis) versus the number of lost convenience transfers with rank "A" ($y$-axis).

transfers and the number of lost important convenience transfers for all time
intervals. All results of optimization runs for workingdays show a similar
structure, where most solutions are located in a corridor that is parallel to
the main diagonal as it is shown in Figure 7.8. In contrast to this, the analyses for Saturday afternoon and for Sunday show a more chaotic structure,
similar to the one of Figure 7.10. Here, changing a transfer so that a different departing tour is chosen from the departing line is not as easy as on
workingdays.

### 7.3.3   Significance of the Results

In this subsection, we examine whether the modeling and the optimization
approaches of this work are able to produce meaningful results for the real
life situation. For this, we analyze several generated timetables in detail.
Note that the goal of the traffic planners is not to find the optimal timetable
in terms of the objective functions. Instead, a well-balanced solution that
does not destroy too many established connections is often favored. The
quality of such a well-balanced timetable cannot be expressed in a single
table; instead the timetables must be regarded as a whole. Nevertheless, we
present two generated solutions: one for the workingday midday timetable
and one for the Saturday evening timetable. In Table 7.2 and Table 7.3, we
visualize the changes to the number of transfers of the different transfer types
for the overall network and the three network nodes. Both solutions refer to
unconstrained problems.

|          | almost      | risk       | conv.        | patience   | no tr.       |
|----------|-------------|------------|--------------|------------|--------------|
| overall  | 519 (-262)  | 221 (-174) | 3007 (+340)  | 1433 (-52) | 5694 (+148)  |
| city ctr | 282 (-252)  | 112 (-165) | 2601 (+357)  | 992 (-85)  | 3999 (+145)  |
| main st. | 215 (-2)    | 100 (+3)   | 388 (-26)    | 417 (+42)  | 1571 (-17)   |
| st. west | 22 (-8)     | 9 (-12)    | 18 (+9)      | 24 (-9)    | 124 (+20)    |

Table 7.2: Number of transfers of the different transfer types for the overall
situation and the three network nodes for an optimized workingday midday
timetable. The solution is achieved by an unconstrained optimization. The
numbers in brackets show the changes that are made to the current timetable.

The tables show significant improvements in the number of convenience
transfers, but the question whether these solutions are acceptable cannot be

|          | almost     | risk      | conv.        | patience  | no tr.       |
|----------|------------|-----------|--------------|-----------|--------------|
| overall  | 325 (-25)  | 190 (+4)  | 2080 (+123)  | 759 (+36) | 6043 (-138)  |
| city ctr | 109 (-64)  | 78 (-39)  | 1675 (+118)  | 405 (+15) | 4169 (-30)   |
| main st. | 194 (+30)  | 101 (+46) | 387 (+9)     | 306 (-3)  | 1738 (-82)   |
| st. west | 22 (+9)    | 11 (-3)   | 18 (-4)      | 48 (+24)  | 136 (-26)    |

Table 7.3: Number of transfers of the different transfer types for the overall situation and the three network nodes for the Saturday afternoon timetable. The solution is achieved by an unconstrained optimization. The brackets show the changes that are made to the current timetable.

answered on the basis of these numbers. A big gain of convenience transfers is nice, but the price for this improvement is unknown. Thus, more information is needed here.

For this, we take a look at the Sunday analysis. The current situation of the transfers that belong to connections that are important (rank "A") is already good synchronized, which can be seen in Table 7.4. Here, the convenience transfers make up the largest part of the transfers that are not "no transfers" (more than 85%). But there is still room for improvement, even for this already good synchronization. Our algorithms are able to generate 39 new convenience transfers for the price of 'only' 11 convenience transfers that are changed into risk or patience transfers. It is remarkable that most of these new gained convenience transfers were almost transfers before. This means that we are able to generate totally new transfer possibilities for the passengers.

A difficult situation in the network of Kaiserslautern is the transfer possibility for the students between the main station and the university. Here, a special bus line (line 116) was introduced to improve the situation. The optimization algorithms propose a small negative shift for line 116 to improve the situation in the evening. The outcome for the transfers between line 116 and a regional train is presented in Figure 7.11. Here, four risk transfers are changed into convenience transfers.

In Table 7.5, we present the consequences of this shift for the situation at the main station. The numbers refer to changes of transfers that belong to important connections (rank "A"). The transfer types are changed 'to the right' in this table, which corresponds to longer waiting times for the passengers.

| from \ to | almost | risk | conv. | patience | no tr. |
|---|---|---|---|---|---|
| almost | 28 | 0 | 23 | 0 | 27 |
| risk | 0 | 31 | 1 | 0 | 0 |
| convenience | 0 | 5 | 1726 | 6 | 0 |
| patience | 0 | 0 | 15 | 206 | 0 |
| no transfer | 0 | 0 | 0 | 0 | 3392 |

Table 7.4: Transfer type changes of the connections with priority "A" for a solution of the Sunday timetable. Unchanged transfers are on the gray diagonal. Only a few changes are made to the current timetable to achieve this solution. There are 39 gained convenience transfers and only 11 lost convenience transfers in the new timetable.

| from \ to | almost | risk | conv. | patience | no tr. |
|---|---|---|---|---|---|
| almost | 68 | 3 | 0 | 0 | 0 |
| risk | 0 | 11 | 10 | 0 | 0 |
| convenience | 0 | 0 | 23 | 8 | 0 |
| patience | 0 | 0 | 1 | 132 | 4 |
| no transfer | 1 | 0 | 0 | 0 | 432 |

Table 7.5: Transfer type changes of the connections with priority "A" at the main station from city buses to trains for the workingday evening timetable. The changes are caused by a shift of $-3$ minutes for line 116. The transfer types are shifted 'to the right', which means longer waiting times for the passengers. The changes are promising, since passengers favor short patience transfers over risk transfers.

A similar situation can be found for the workingday morning timetable. But in contrast to the evening timetable, the proposed shift for line 116 is here $+2$ minutes. Since the students transfer in the morning from the train lines to the bus line, this shift generates longer transfer times and the comfort for the passengers is increased.

These two examples show that our concept is able to find suboptimal situations in real life timetables. Here, new solution strategies are generated automatically to improve the situation significantly.

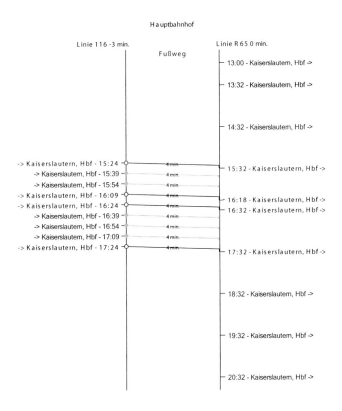

Figure 7.11: Improvement of the situation for the students traveling from the university to the main station in the evening. Four "risk transfers" (marked by the unfilled dots) are changed into "convenience transfers". The black diagonal lines represent "convenience transfers", the gray lines show "no transfers".

## 7.4  Summary of the Chapter

In this chapter, we regard an application of the approaches of the previous chapters to a real life situation. The traffic network is of medium size and contains three different public transport companies. For this problem size, the metaheuristic approaches are able to generate good solutions in a reasonable amount of time.

Our solution approach is able to detect problematic situations in the real life timetables. Here, solution strategies are generated automatically to improve the transfer possibilities significantly. This shows that our modeling approach reflects the real life situation in a good way. The different constraints enforce that the generated solutions fulfill the goals of the traffic planner.

Within the set of the generated solutions, there are several timetables that have a good ratio between the number of generated convenience transfers of connections with rank "A" and the number of lost convenience transfers of connections with rank "A". Especially these solutions are interesting for the traffic planners, since the timetables are locally improved without destroying too many important transfer possibilities.

# Chapter 8

# Conclusion and Outlook

In the final chapter of this thesis, we present a conclusion of the main results of this work. Furthermore, an outlook on future research and possible applications of our approach is given.

## 8.1 Public Transport Timetable Synchronization

In Chapter 2, we presented a new approach for modeling the TTSP. With this approach, we are able to generate meaningful timetables that offer an enhanced synchronization at the intersections of the networks of different public transport companies. Traffic planners can adjust the model and control the optimization by placing constraints. Here, we introduce an interactive process between the traffic planners and the mathematical algorithm, where the traffic planner analyzes the computed solutions and can place constraints to direct the algorithm so that it finds solutions in the next iteration that fit his needs better.

The model is focused on passenger convenience. Our goal is to find a good compromise between having short waiting times and minimizing the risk of missing a transfer due to delays. The timetables are analyzed with respect to the trade-off between the local and the global character of the problem. Each single transfer possibility represents an optimization goal of its own; thus, the multicriterian aspects of the problem must be regarded as well. This is taken into account in our approach by analyzing the network on different levels of detail and by weighting the different transfers due to their importance.

We model the TTSP as a mathematical optimization problem (QSAP) and compute solutions with different metaheuristics (Ant Colony Optimization, Genetic Algorithm and Simulated Annealing). We present the ACO approach in detail and introduce a new way to deal with the multicriterian character of the problem by using several ant colonies simultaneously. The quality of the generated solutions is evaluated with different approaches that are able to generate lower bounds for the QSAP. The metaheuristics generate Pareto optimal solutions in an acceptable time, even for real life test instances. This is achieved by hybridizing the approaches with a local search heuristic that further improves the generated solutions.

The model is applied to the real life public transport network of the medium-sized city of Kaiserslautern. Here, the optimization approaches are able to find and improve suboptimal transfer situations in the traffic network. For the overall situation, improvements of more than six percent can be achieved.

## 8.2   Mathematical Theory

The QSAP used in our research has a more general form compared to the one that is generally used in literature. The new aspect is that we allow different sets of locations for the single objects that must be assigned. For this new type of QSAP, we have extended the known polyhedral results so that they include these flexible location sets.

We present an MIP formulation that is able to calculate an optimal solution for a single criterion QSAP. But the computation of this optimum takes far too much time for real life problem sizes, even if only one objective function is regarded. Therefore, we use lower bound approaches to obtain an estimation of the optimal solution value of our problem.

Trivial lower bound approaches are able to calculate bounds quickly, but the gaps of these bounds are far too large to be useful for our problem. Additionally, the quality of these lower bound approaches depends to a large degree on the structure of the objective function. Therefore, better and more stable lower bounds are needed.

A good lower bound approach is the Reformulation Linearization Technique that is known to generate promising results for the QAP. The RLT provides stepwise tightening polyhedral formulations of the QSAP. We extend the mathematical theory of the RLT for QSAPs by proving that a certain RLT level provides an exact polyhedral characterization of the QSAP. This result is achieved by finding minimal problem structures that cause untight solutions. For most QSAPs, this value improves the formerly known trivial tight formulation RLT-($m$-1). With this knowledge, we construct a new algorithmic approach that breaks up the strict level structure of the RLT. This RLT-1.5 algorithm improves the RLT-1 formulation stepwise by integrating RLT-2 constraints to eliminate occurrences of this minimal problem structure.

A different approach that we use to generate lower bounds for QSAPs is to approximate the height of the objective function of the QSAP. Here, we extend the standard definition of the height by introducing a stepwise improving formulation. Our Dual Ascent algorithm (DA2) approximates the height of degree three by extracting constant terms of the objective function so that the remaining posiform is a cubic function. The algorithm does not need any external LP or MIP solvers and it is able to generate good bounds quickly.

The concept of approximating the height of the objective function is closely related to the ideas of the RLT formulation. Both theories result in stepwise improving lower bound approaches that become tight from a certain level on. Therefore, we transfer the idea of the RLT-1.5 approach to our Dual Ascent approach. The new concept that we call a Dual Ascent 2.5 approach shows promising results.

The best bounds for the QSAP are achieved by a hybrid algorithm that combines the Dual Ascent level 2 approach with the RLT-1.5 approach. Here, the fast utilization of the problem structure by the Dual Ascent approach is combined with the analytical strength of the RLT-1.5 approach.

## 8.3  Outlook

We have successfully applied the presented approach for synchronizing timetables in public transport to a real life situation in a medium-sized city. Nevertheless, there is still room for improvement of the concept. Up to now, our work has the focus on pre-existing timetables. But the concept also has a high potential regarding the initial creation process of timetables. In most cases, this process is based on the given timetables of the previous timetable period. These old timetables are modified to fit the needs of the new period, which can result from changed passenger flows or because one company has changed its timetable and the other companies need to adjust their timetables to preserve the present transfers.

This timetable creation process is performed manually at the moment and there is not much time to make any necessary adjustments if one company changes its timetable near the deadline for the completion of the network-wide timetable. Our concept of supervising the impact of changes made to the system and the possibility to generate good solutions automatically may help to improve this process significantly.

For the Reformulation Linearization Technique, an interesting future research topic is to transfer the theory of the minimal problem graphs $G_{RLT}^t$ to the Quadratic Assignment Problem. A first result in this area is presented in Subsection 5.6.4, but the problem of finding the smallest RLT level $t$ for a certain QAP size so that the RLT-$t$ formulation is tight still remains unsolved. For this, a general minimal problem graph structure $\widetilde{G}_{RLT}^t$ is needed, similar to our approach for the QSAP.

Another topic that can be addressed in future research is the connection between the general definition of the height $H_{t,\mathcal{A}}[f(x)]$ of the objective function $f$ of a QSAP and the RLT-$t$ formulation of the same QSAP. Both, the height approach and the RLT approach, generate a stepwise improving series of lower bounds for the QSAP. It is known that

$$H_{2,\mathcal{A}}[f(x)] = \min \ \text{RLT-1}$$

holds for any QSAP and we have proven in addition that

$$H_{m,\mathcal{A}}[f(x)] = \min \ \text{RLT-(m-1)} = \min \ \text{QSAP}$$

holds for all QSAPs with $m$ rows. Whether the equation

$$H_{t+1,\mathcal{A}}[f(x)] = \min \ \text{RLT-}t$$

holds for all $t \in \{1, ..., m-1\}$ is still unknown for QSAPs. The results from Section 5.6 and Section 5.8 may help to answer this question in future research.

For the hybrid approach that we present in this thesis, a higher level of Semi-Assignment Problems can be useful. The transformed objective functions of the Dual Ascent level 2 algorithm contain cubic terms. Hence, the theory of Cubic Semi-Assignment Problems may be of help to improve our hybrid approach. Here, the Reformulation Linearization Technique is again a promising concept, since its linearization technique also works on cubic objective functions.

# List of Symbols

| | |
|---|---|
| $\omega$ | Station |
| $\omega_{arr}$ | Arriving station |
| $\omega_{dep}$ | Departing station |
| $\Omega$ | Set of all stations |
| $\delta$ | Network node |
| $\Delta$ | Set of all network nodes |
| $\zeta$ | Tour |
| $\zeta_{arr}$ | Arriving tour |
| $\zeta_{dep}$ | Departing tour |
| $T$ | Set of all tours |
| $t_{arr}$ | Arriving time |
| $t_{dep}$ | Departing time |
| $l$ | Line |
| $l_{arr}$ | Arriving line |
| $l_{dep}$ | Departing line |
| $L$ | Set of all lines |
| $\lambda$ | Line cluster |
| $\Lambda$ | Set of all line clusters |
| $\rho$ | Direction |
| $c$ | Connection |
| $C$ | Set of all connections |
| $\tau$ | Transfer |
| $\mathcal{T}$ | Set of all transfers |
| $t_{trans}$ | Transfer time |
| $t_{drive}$ | Driving time |
| $t_{walk}$ | Walking time |
| $t_{wait}$ | Waiting time |
| $\mathcal{W}$ | Walking time function for a pair of stations and lineclusters |
| $\xi$ | Waiting time function for a pair of tours |
| $P$ | Penalty function |
| $\widehat{P}$ | Smoothed penalty function |
| $s$ | Shift of a line |
| $S$ | All possible shifts of a line |
| $A$ | Assignment |

| | |
|---|---|
| $\mathcal{A}$ | Set of all assignments |
| $t$ | Level of the RLT-formulation |
| $\sigma$ | RLT feasible solution |
| $\phi$ | Posiform |

# Bibliography

[1] W. P. Adams, M. Guignard, P. M. Hahn, and W. L. Hightower. A level-2 reformulation-linearization technique bound for the quadratic assignment problem. *European Journal of Operational Research*, 180(3):983–996, August 2007.

[2] A. Adamski and Z. Bryniarska. Schedule synchronization in public transport by tabu search and genetic method, 1997.

[3] Bahnaktuell. Vvo stimmt Anschlüsse besser ab. `http://www.bahnaktuell.net/BA2/wordpress/?p=4313`, December 2008.

[4] A. Billionnet, M. C. Costa, and A. Sutter. An efficient algorithm for a task allocation problem. *J. ACM*, 39(3):502–518, 1992.

[5] A. Billionnet and S. Elloumi. Best reduction of the quadratic semi-assignment problem. *Discrete Applied Mathematics*, 109(3):197–213, 2001.

[6] C. Blum and A. Roli. Metaheuristics in combinatorial optimization: Overview and conceptual comparison. *ACM Computing Surveys*, 35(3):268–308, 2003.

[7] J. H. Bookbinder and A. Désilets. Transfer optimization in a transit network. *Transportation science*, 26(2):106–118, 1992.

[8] E. Boros and P. L. Hammer. Pseudo-boolean optimization. *Discrete Applied Mathematics*, 123(1-3):155–225, 2002.

[9] S. Burer and D. Vandenbusschue. Solving lift-and-project relaxations of binary integer programs. *SIAM Journal on Optimization*, pages 726–750, 2006.

[10] R. E. Burkard, S. E. Karisch, and F. Rendl. QAPLIB - a quadratic assignment problem library, 1996.

[11] A. Désilets and J.-M. Rousseau. Syncro: A computer-assisted tool for the synchronization of transfers in public transit networks. In *Computer-Aided Transit Scheduling: Proceedings of the Fifth International Workshop on Computer-Aided Scheduling of Public Transport*, volume 386 of *Lecture Notes in Economics and Mathematical Systems*, pages 153–166, August 1990.

[12] Y. Ding and H. Wolkowicz. A low dimensional semidefinite relaxation for the quadratic assignment problem, 2009.

[13] W. Domschke. Schedule synchronization for public transit networks. *OR Spectrum*, 11(1):17–24, 1989.

[14] M. Dorigo and C. Blum. Ant colony optimization theory: A survey. *Theoretical Computer Science*, 344(2-3):243–278, 2005.

[15] M. Dorigo and T. Stützle. *Ant Colony Optimization*. The MIT Press, Cambridge, Massachusetts, 2004.

[16] C. Fleurent, R. Lessard, and L. Séguin. Transit timetable synchronization: Evaluation and optimization. *9th International Conference on Computer Aided Scheduling in Public Transport (CASPT), San Diego, California, August*, pages 9–11, 2004.

[17] C. M. Fonseca and P. J. Fleming. Genetic algorithms for multiobjective optimization: Formulation, discussion and generalization. In *Genetic Algorithms: Proceedings of the Fifth International Conference*, pages 416–423. Morgan Kaufmann, 1993.

[18] C. M. Fonseca and P. J. Fleming. An overview of evolutionary algorithms in multiobjective optimization. *Evolutionary Computation*, 3(1):1–16, 1995.

[19] H.K. Fung, S. S. Rao, C. A. Floudas, O. A. Prokopyev, P. M. Pardalos, and F. Rendl. Computational comparison studies of quadratic assignment like formulations for the in silico sequence selection problem in de novo protein design. *Journal of Combinatorial Optimization*, 10:41–60, 2005.

[20] S. Goss, S. Aron, J. L. Deneubourg, and J. M. Pasteels. Self-organized shortcuts in the argentine ant. *Naturwissenschaften*, 76(12):579–581, 1989.

[21] T. Grünert, S. Irnich, H.-J. Zimmermann, M. Schneider, and B. Wulfhorst. Finding all k-cliques in k-partite graphs, an application in textile engineering. *Computers and Operations Research*, 29(1):13–31, January 2002.

[22] P. M. Hahn, W. L. Hightower, T. A. Johnson, M. Guignard, and C. Roucairol. A lower bound for the quadratic assignment problem based on a level-2 reformulation- linearization technique. *Systems Engineering Department Report, University of Pennsylvania*, 2001.

[23] P. M. Hahn, B.-J. Kim, M. Guignard, J. M. Smith, and Y.-R. Zhu. An algorithm for the generalized quadratic assignment problem. *Comput. Optim. Appl.*, 40(3):351–372, 2008.

[24] P. M. Hahn, Y.-R. Zhu, M. Guignard, and W. L. Hightower. A level-3 reformulation-linearization technique bound for the quadratic assignment problem. *Optimization Online*, 2008.

[25] P. L. Hammer, P. Hansen, and B. Simeone. Roof duality, complementation and persistency in quadratic 01 optimization. *Mathematical Programming*, 28(2):121–155, 1984.

[26] N. Hansen. Integrating timetabling and vehicle scheduling. diploma thesis, Technical University, Kaiserslautern, May 2009.

[27] A. Jaszkiewicz. Multiple objective metaheuristic algorithms for combinatorial optimization. Habilitation Thesis, Poznan University of Technology, 2001.

[28] M. Jünger and V. Kaibel. On the SQAP-polytope. *SIAM J. Optimization*, 11(2):444–463, 2000.

[29] M. Jünger and V. Kaibel. The QAP-polytope and the star-transformation. *Discrete Appl. Math.*, 111:283–306, August 2001.

[30] M. Khichane, P. Albert, and C. Solnon. Cp with aco. *CPAIOR 2008*, pages 328–332, 2008.

[31] S. Kirkpatrick, C. D. Gelatt Jr., and M. P. Vecchi. Optimization by simulated annealing. *Science*, 220(4598), May 1983.

[32] S. Klein. Multicriteria optimization methods for schedule synchronization in public transit. Master's thesis, TU Kaiserslautern, 2006.

[33] W. Klemt and W. Stemme. Schedule synchronization for public transit networks. In J. R. Daduna and A. Wren, editors, *Computer-aided transit scheduling. Proceedings, Hamburg, Germany, 1987. Berlin: Springer. Lect. Notes Econ. Math. Syst.*, volume 308, pages 327–335, 1988.

[34] C.-G. Lee and Z. Ma. The generalized quadratic assignment problem. *Research Report, Department of Mechanical and Industrial Engineering, University of Toronto, Toronto, Ontario, M5S 3G8, Canada*, 2004.

[35] C. Liebchen. Der Berliner U-Bahn Fahrplan 2005 Realisierung eines mathematisch optimierten Angebotskonzepts. In *Heureka '05 Optimierung in Transport und Verkehr*, pages 483–500. FGSV Verlag, Koln, Germany, 2005.

[36] C. Liebchen. *Periodic Timetable Optimization in Public Transport*. dissertation.de, 2006.

[37] M. López-Ibáñez, L. Paquete, and T. Stützle. On the design of aco for the biobjective quadratic assignment problem. In Marco Dorigo, Mauro Birattari, Christian Blum, Luca Maria Gambardella, Francesco Mondada, and Thomas Stützle, editors, *ANTS Workshop*, volume 3172, pages 214–225. Springer, 2004.

[38] F. Malucelli. A polynomially solvable class of quadratic semi-assignment problems. *European Journal of Operational Research*, 91(3):619–622, June 1996.

[39] F. Malucelli and D. Pretolani. Lower bounds for the quadratic semi-assignment problem. *European Journal of Operational Research*, 83(2):365–375, 1995.

[40] D. Merkle, M. Middendorf, and S. Iredi. Bi-criterion optimization with multi-colony ant algorithms. In Eckart Zitzler, Kalyanmoy Deb, Carlos A. Coello Coello, and David Corne, editors, *First International Conference on Evolutionary Multi-Criterion Optimization*, volume 1993 of

*Lecture Notes in Computer Science*, pages 359–372. Springer Verlag, Berlin, 2001.

[41] M. Middendorf and D. Merkle. Modeling the dynamics of ant colony optimization. *Evolutionary Computation*, 10(3):235–262, 2002.

[42] G. L. Nemhauser and L. A. Wolsey. *Integer and Combinatorial Optimization*. Wiley-Interscience, New York, 1988.

[43] A. A. Pessoa, P. M. Hahn, M. Guignard, and Y.-R. Zhu. An improved algorithm for the generalized quadratic assignment problem. *Simpósio Brasileiro de Pesquisa Operacional, 2008, João Pessoa. Anais do XL SBPO*, 2008.

[44] N. A. Pierce and E. Winfree. Protein design is NP-hard. *Protein Eng.*, 15(10):779–782, October 2002.

[45] A. Radev. Timetable synchronization in public transport: Modeling planner preferences and optimization with multiobjective evolutionary algorithms. Master's thesis, Universität Kaiserslautern, 2008.

[46] F. Rendl and R. Sotirov. Bounds for the quadratic assignment problem using the bundle method. *Mathematical Programming*, 109(2-3):505–524, 2007.

[47] F. Roupin. From linear to semidefinite programming: an algorithm to obtain semidefinite relaxations for bivalent quadratic problems. *Journal of Combinatorial Optimization*, 8(4):469–493, 2004.

[48] H. Saito. The symmetric quadratic semi-assignment polytope. *IEICE Transactions*, 89-A(5):1227–1232, 2006.

[49] H. Saito, T. Fujie, T. Matsui, and S. Matuura. The quadratic semi-assignment polytope. Mathematical Engineering Technical Reports METR 2004-32, University of Tokyo, 2004.

[50] A. Schöbel. *Optimization in Public Transportation*. Optimization and Its Applications. Springer, New York, 2006.

[51] M. Schröder and I. Schüle. Interaktive mehrkriterielle Optimierung für die regionale Fahrplanabstimmung in Verkehrsverbünden. *Straßenverkehrstechnik*, (6):332–340, June 2008.

[52] M. Schröder and I. Solchenbach. Optimization of transfer quality in regional public transit. Technical report, Fraunhofer ITWM, 2006.

[53] I. Schüle, A. Dragan, A. Radev, M. Schröder, and K.-H. Küfer. Multi-criteria optimization for regional timetable synchronization in public transport. In *Operations Research Proceedings 2008*, pages 313–318. Springer Verlag, September 2008.

[54] I. Schüle, H. Ewe, and K.-H. Küfer. Finding tight RLT formulations for quadratic semi-assignment problems. In *CTW*, pages 109–112, 2009.

[55] I. Schüle, M. Schröder, and K.-H. Küfer. Synchronization of regional public transport systems. In *Urban Transport XV*, June 2009.

[56] H. D. Sherali and W. P. Adams. *A Reformulation-Linearization Technique for Solving Discrete and Continuous Nonconvex Problems*, volume 31 of *Series: Nonconvex Optimization and Its Applications*. Springer,, 1998.

[57] H. D. Sherali and L. Liberti. *Reformulation-Linearization Methods for Global Optimization*, pages 3263–3268. Springer, Berlin, 2008.

[58] J. Sohn and S. Park. The single allocation problem in the interacting three-hub network. *Networks*, 35(1):17–25, December 1999.

[59] C. Solnon. Ants can solve constraint satisfaction problems. *IEEE Transactions on evolutionary computation*, 6(4):347–357, August 2002.

[60] T. Stützle. *Local Search Algorithms for Combinatorial Problems - Analysis, Improvements, and New Applications*. PhD thesis, Technische Universität Darmstadt, 1999.

[61] T. Stützle and H. H. Hoos. Max-min ant system. *Meta-Heuristics: Advances and Trends in Local Search Paradigms for Optimization*, pages 137–154, 1999.

[62] S. Voß. Network design formulations in schedule synchronization. *Computer-aided transit scheduling. Procédings, Montreal, Canada, Berlin: Springer. Lect. Notes Econ. Math. Syst.*, pages 137–152, August 1992.

[63] H. Wolkowicz. *Semidefinite Programming Approaches To The Quadratic Assignment Problem*, pages 143–174. Kluwer Academic Publishers, 2000.

[64] Q. Zhao, S. E. Karisch, F. Rendl, and H. Wolkowicz. Semidefinite programming relaxations for the quadratic assignment problem. *Journal of Combinatorial Optimization*, 2(1):71–109, March 1998.

[65] Y.-R. Zhu. *Recent Advances and Challenges in Quadratic Assignment and Related Problems*. PhD thesis, University of Pennsylvania, 2007.

# The Authors Scientific Career

## Studies

Oct. 2007 - Feb. 2010    PhD study in Mathematics at the Technical University of Kaiserslautern (funded by a scholarship of the Fraunhofer- Institut für Techno- und Wirtschaftsmathematik (ITWM)).

Jan. 2007 - Sep. 2007    Project Studies in Advanced Technologies (ProSat) at the Technical University of Kaiserslautern (funded by a scholarship of the Fraunhofer- Institut für Techno- und Wirtschaftsmathematik (ITWM)).

Aug. 2004 - Jan. 2005    Study in Mathematics at the Royal Technical University Stockholm (KTH)

Oct. 2001 - Dec. 2006    Diploma study in Mathematics at the Technical University Clausthal.

        Title of the diploma thesis:
        *Klassifizierung perfekter Codes.*

# Publications

The following publications contain parts of this thesis or contain precursory work:

[51] Schröder, M. and Schüle, I. Interaktive mehrkriterielle Optimierung für die regionale Fahrplanabstimmung in Verkehrsverbünden, Straßenverkehrstechnik:332-340, 2008.

[53] Schüle, I. and Dragan, A. and Radev, A. and Schröder, M. and Küfer, K. -H. Multi-criteria optimization for regional timetable synchronization in public transport, Operations Research Proceedings 2008:313-318, 2008.

[54] Schüle, I. and Ewe, H. and Küfer, K. -H. Finding Tight RLT Formulations for Quadratic Semi-Assignment Problems, CTW:109-112, 2009.

[55] Schüle, I. and Schröder, M. and Küfer, K. -H. Synchronization of regional public transport systems. Urban Transport XV, 2009.